Managing Creativity in Science and Hi-Tech

Ronald Kay

Managing Creativity in Science and Hi-Tech

Second Edition

 Springer

Ronald Kay
San Francisco CA, USA
nukayron@mac.com

ISBN 978-3-642-44140-0   ISBN 978-3-642-24635-7 (eBook)
DOI 10.1007/978-3-642-24635-7
Springer Heidelberg Dordrecht London New York

Printed on acid-free paper

Springer is part of Springer Science+Business Media (www.springer.com)

*To Renate, whose insight
and commitment has given
balance to our joint ventures*

Preface

This new edition was motivated by the realization that the issues addressed have become more significant and more widespread. The importance of Scientific and Hi-Tech enterprise has increased the world over. The need for people with scientific and technical competence has become ubiquitous, including the highest levels of management in industry, education, and government. Today, this need extends to every organization which depends upon information and communication technology, that is, not only to science and high tech but also to banking, finance and insurance, marketing, and the energy industry.

During the past 20 years, Hi-Tech enterprises have become the most dynamic segment of the world's economy. Information processing and communication technology in particular have changed the world we live in. Very likely, Bio-Hi-Tech will become an equally significant factor in the course of the twenty-first century.

Industrial research facilities have played a dominant part, by accepting (sometimes reluctantly) the role of advance troops to the development army which turns invention into innovation. University departments which address the issues germane to high technology have made major contributions and have found effective ways of interacting with industry.

This is hardly surprising given the magnitude of the effort. Worldwide R&D expenditures in 2011 were in excess of $1,000 billion ($10^{12}$).

Much of that R&D effort has assumed international dimensions. Large companies like Siemens, NEC, Novartis, IBM, and Microsoft have laboratories all over the world. The cost of introducing new technologies has motivated companies to find joint venture partners in every part of the world. The complexity and scope of management issues which have accompanied this development is making new demands upon people whose formal education has been largely concentrated upon science and engineering.

There has not been a commensurate advance in the training of engineers and scientists to prepare for management responsibility.

Higher education has only begun to meet this need. An advanced scientific or technical education still leaves little time to prepare the most capable for management responsibility which they are sure to face.

To fill the need at the top, it is essential to motivate and help scientists and engineers meet management challenges at every stage of their career.

This new edition includes a look at the international aspects of the evolving environment. This is a matter of particular urgency.

Europe, Japan, and the USA will face competition from China and India in the immediate future. The latter have invested heavily in the education of potentially creative individuals. They have the benefit of people who have experienced the best training that Europe and the US have to offer.

These new Asian leaders in Science and Hi-Tech have an advantage which is hard to overstate: They have come to accept, that there is another way of doing things, the hallmark of creativity. They have had the experience of adapting to another culture and selecting that which promises to serve *their future* needs.

This new edition also considers the role of technology itself upon the challenges and opportunities faced by managers of Science and Hi-Tech.

It has been reassuring that recent studies of the effectiveness of management in science and high-tech have borne out the continuing relevance of what has been put forth in the earlier edition. Google's Project Oxygen (2011), a data driven survey of employee views of management effectiveness, is an especially relevant validation.

San Francisco, California   *Ronald Kay*
February 2012

Acknowledgments

It has been my good fortune to be able to call upon people who have played significant roles in the advancement of the management of Science and Hi-Tech since the publication of the earlier edition.

Of the many people consulted, I am especially indebted to those who have been so generous with their views and their time.

Clemens Szyperski, who has made the journey from successful Hi-Tech entrepreneur in Switzerland to Academia in Australia, advanced the science and practice of software engineering and is a manager and leading software architect at Microsoft in Redmond, Washington.

Donald Shapero, Director, Board on Physics and Astronomy of the US National Academy in Washington, DC, with a lifetime in national science policy.

Evgeny Zaytsev of Helix Ventures who made the journey from managing a major Russian biomedical program in Siberia to starting a Venture Capital firm in Palo Alto, California.

Frank Mayadas, who has held executive positions at IBM and with the Alfred P. Sloan Foundation, New York, spend a lifetime in Science and Hi-Tech management and was instrumental in the advancement of E-Learning.

Hans Danielmeyer, Stuttgart, Germany, whom I first met at a Management Workshop for young physicists. He has since been involved with every aspect of Science and Hi-Tech management as founding President of a Graduate Technical University, as Board Member of Siemens, and as member of German and European Commissions

Hervé Bourlard, Founding Director of the IDIAP Research Institute in Martigny, Switzerland, singled out for Entrepreneurship and recognized for his contributions to the field of Speech Processing and Artificial Neural Networks.

Moidin Mohiuddin, Associate Director of the IBM Research Laboratory in Almaden, California, who helped me put some of the changes over the past 20 years in perspective.

Norbert Szyperski, University of Cologne and Sylter Runde, who has held CEO positions in public and private Hi-Tech organizations in Germany and has played a leading role in establishing and in making the case for Hi-Tech start-ups in Europe.

Thomas Huber, Switzerland, could draw upon his unique experience, going from the University of Zurich Biochemistry Department directly into a management job with the Novartis Pharma Research Lab.

Uwe Thomas, Friedrich-Ebert Stiftung, Germany, as director at the Federal Ministry for Research and Technology played a decisive role in creating the conditions for Germany's re-emergence in Science and Hi-Tech.

Wolfgang Wahlster, Director of the DFKI, Artificial Intelligence Centers in Germany and advisor to government and industry in various parts of the world has a comprehensive view of the challenges facing Science and Hi-Tech management.

Preface to the First Edition

The growing role of science and technology has generated a need for unique management skills on the part of scientists and engineers. While this need is widely recognized, there is little agreement on the most appropriate way in which it should be satisfied.

By and large, the general literature on management does not recognize problems that are unique to science and high technology. This lack is also reflected in the considerable variety of formal management training. More often than not it has missed its mark, at least when judged by the response of participating scientists and engineers.

The principle motivation for this book has been my own experience and teaching graduate students and practicing scientists and engineers about those aspects of management that are likely to be most relevant to their future endeavors.

The book reflects some of what I have learned from that experience and has been further encouraged by the convictions that

1. The distribution of management potential among engineers and scientists is no different from that of other groups with comparable academic achievement.
2. Successfully managed scientific and technical organizations provide the most useful source of learning.
3. The process of learning is facilitated by referring to the experience that has proven effective in creating an environment in which creative scientific and technical enterprise has flourished.

I have found such flourishing enterprise in universities, in government laboratories, and in Hi-Tech industry, the world over.

Creativity plays a special role in such enterprise. It is the recognition of that special role which makes unique demands upon those who assume management responsibility. Management principles that have been applied successfully to administration, finance, marketing, and production need modification when applied to creative scientific and technical effort.

Every time an individual scientist or engineer selects a problem to be tackled or an approach to its solution, she or he assumes personal risk, often with long-term implications. This readiness to assume risk is a necessary, if not sufficient, condition for creativity to flourish. Hence, the environment for creative endeavor must be conducive to risk taking.

Besides bringing relevant experience to bear upon the topic of managing creativity, this book is also intended to convey some of the challenge and excitement which awaits those willing to assume responsibility.

October 1989   *Ronald Kay*

Acknowledgments to the First Edition

It has been my good fortune to have had close association with a number of very effective managers, and to have spent a good part of my professional life with the International Business Machines Corporation. IBM is a company which places a premium upon creativity and works hard at the task of developing managers.

For many years, daily discussions of my wife's experiences, managing a class of 30 ten-year-olds, has helped me put management in perspective: To help people develop their potential.

From my children, I have learned to recognize the limitations of age-induced one-upmanship.

In life, one owes a lot to those who have put up with one's failings, a most valuable source of learning. To all of them, my sincere gratitude; in particular, to Angela Lahee, my most effective editor at Springer-Verlag.

Contents

1. Introduction

Creativity is the most precious asset engineers and scientists bring to the Hi-Tech environment. The nurturing of this asset has made the difference in the long-term competitiveness of nations and enterprises.

By its very nature, Hi-Tech enterprise requires creative people with a strong technical background to manage not only research and development, but also other functions such as administration, manufacturing, marketing, and sales.

Beyond Hi-Tech, the need for highly qualified and creative technical people in management extends to organizations concerned with environmental issues, to government, and to industries dependent on information technology, such as finance, insurance, and the like; in short, wherever technology plays an important part.

The global implications of Science and Hi-Tech have added another dimension to the demands made upon those who have or will take on management responsibility.

Traditionally, the organization chart has defined the management hierarchy as well as the decision-making process. But in organizations where information and rapidly changing know-how play a significant role, up-to-date technical knowledge must be introduced into the decision-making process at all levels. As a consequence, the most capable and creative scientists and engineers are asked to assume management responsibility very early in their careers.

In the past, such responsibility was not thrust upon them until they had acquired some "experience." Now, however, there simply are not enough "experienced" people available who have an adequate, state-of-the-art, technical background.

In this environment, the starting professional and the newly appointed manager face a unique problem: the opportunity of learning from experience is not sufficient to make up for the almost total lack of preparation for the nontechnical aspects of management during the course of a formal technical/scientific education.

This problem can be further aggravated by a negative attitude toward the very concept of management on the part of people who have acquired an advanced degree. Such an attitude is not uncommon in graduate science and engineering departments in many universities. Aversion to management responsibility on the part of university faculty is generally recognized. In scientific and engineering graduate

R. Kay, *Managing Creativity in Science and Hi-Tech*,
DOI 10.1007/978-3-642-24635-7_1, © Springer-Verlag Berlin Heidelberg 2012

schools, there is often an additional element of elitism toward management-related academic pursuit and thus toward management itself.

The "academic atmosphere" of some graduate departments encourages a negative or even disparaging attitude toward management on the part of young scientists and engineers. Yet, management responsibility is thrust upon the most promising people emerging from that environment.

There are, of course, notable exceptions to this image of science and engineering faculty. Some universities encourage faculty consultation to industry and government, an activity bound to develop an appreciation for management issues. "At Stanford University there was a joke that faculty members could not get tenure until they started a company" (Levy 2011). Some schools encourage an additional MBA degree which goes a long way toward teaching the nontechnical skills of management.

> But a significant portion of the best-trained scientists and engineers, destined to become managers within 5 years after ending their formal education, are ill equipped to assume management responsibility. In spite of 16–20 or more years of education, scientists and engineers find themselves without an adequate frame of reference in dealing with the nontechnical aspects of management.

What is more, many of the most highly valued engineers and scientists have personal characteristics which are at odds with the need for positive interpersonal relationships at the work place. These characteristics may be the result of natural inclination or may be acquired or developed as a way of adapting to a unique environment. A negative attitude toward management on the part of creative professionals may be reinforced by an early experience with a newly appointed and inexperienced manager. Very likely, this manager has come from the ranks of scientists or engineers and was selected for creative ability, rather than interpersonal skills.

This book about the Management of Creative People is intended to question and hopefully influence some of the attitudes toward the subject of management, prevalent among engineers and scientists. It also endeavors to offer a new understanding to the experienced manager who must deal with creative people.

The book's principle objective is to bring about an awareness of the most significant issues facing managers dealing with creative people, and to examine some of the ways managers have learned to deal with these issues. The material is based, to a large extent, upon personal experience. It draws also upon management training material, which I have found particularly relevant and effective in dealing with creative people in a variety of situations, be it in an R&D laboratory, a university, or a new Hi-Tech venture.

The book also points out some relevant literature that has brought the art of management toward a scientific discipline.

The discipline of management lacks some of the unity and rigor of the physical sciences. In that sense, it may be compared to medicine.

Nevertheless, in the case of management as with medicine, the results can provide dramatic evidence of the skills of the practitioners.

The experienced manager will be reassured by the forthright discussion of some of the most difficult, personal aspects of management which are all too often avoided.

Room has been left for divergent views about some of the more controversial aspects of the subject. On the other hand, I have not hesitated to take a position when it seemed appropriate.

2. Outline

Exploration of the question "Is Management of Creative People Desirable?"—from the point of view of the creative individual and from that of the experienced management professional (Sect. 3.1)—establishes a hypothesis to be challenged and tested by what is to follow. The "Characteristics of a Creative Professional" identifies some of the unique problems scientists and engineers face in assuming management responsibility (Sect. 3.3).

The value of both individual and team effort in the pursuit of creative work is recognized (Sects. 3.5 and 3.6).

Rather than attempting a formal definition of "Management," we consider the various ways creative people in general—that is, not only managers—respond to common problems such as the following:

- How we manage our own work
- How we manage our time (Sect. 4.6)
- How we manage our relationship with others somehow involved with our work; for example, how we deal with a negatively perceived manager (Sect. 4.3)

Next, we look at management as it relates to the professional contemplating her/his first management assignments or to the newly appointed manager.

- How do experienced managers select new managers? (Sect. 5.1)
- What are the criteria for selecting people for entry-level, mid-level, and top-level management? (Sects. 5.2–5.4)
- What differentiates the responsibilities at the various levels?

The first encounter with management responsibility at the project level focuses on the sudden change in perception by, and about, the newly appointed manager (Sect. 6.1). The value of schedules, reports, and patent-related documentation in meeting management responsibilities aims to put a different slant on "paper work," as generally perceived (Sects. 6.3, 6.7, and 6.8).

At the next level of management, additional responsibilities must be met. The development of a strategy, operating and financial plan, and their importance to

R. Kay, *Managing Creativity in Science and Hi-Tech*,
DOI 10.1007/978-3-642-24635-7_2, © Springer-Verlag Berlin Heidelberg 2012

large and small organizations is considered, and the elements of a written strategy
and operating plan are identified (Sects. 7.1–7.4).

Throughout the preceding discussion, the management of people has appeared
as a common thread. At this point, we focus upon this most important topic in
a discussion of recruiting, performance evaluation, and compensation (Sects. 8.1–
8.3). In Sect. 8.4, we address the question "Is there *a way* of managing people,
particularly creative, highly motivated people?"

An issue of singular importance is the evaluation of R&D projects (Chap. 9).
Unique to a creative environment is the level of technical knowledge the manager
must bring to the evaluation of ongoing projects as well as to the evaluation of
proposals for the pursuit of new ideas.

- What considerations other than technical are relevant to such evaluations?
- What is relevant to the evaluation of an R&D function and what are the
  appropriate criteria?

Among the required administrative skills, making presentations (Sect. 10.1) and
conducting meetings (Sect. 10.2) stand out as the most demanding in this environment. These skills are equally valuable to the nonmanagerial professional.

Chapter 11 devoted to starting a Hi-Tech enterprise is relevant to both the intra-
and the entrepreneur. We focus upon the aspects of starting a new enterprise which
are unique or particularly important in the Hi-Tech environment or to the potential
entrepreneur with a technical/scientific background. The business plan summary and
the discussion of financial controls emphasize the importance of these topics.

The Venture Capital Industry's impact upon the financing and management of
creative enterprises has been generally recognized if not always well understood
(Sect. 12.1). Much is to be learned from that industry which has broad applicability
to the management of creative effort.

Charitable Foundations' funding plays a unique role along with government
agencies which support R&D with grants (Sects. 12.2 and 12.3).

To nurture creativity, there is need for a set of principles which provide
direction without imposing unwanted constraints. Such principles evolve from the
organizational culture and are an important issue for top management (Chap. 13).
The topics of trust, judgment, and values, generally evaded, are here met head-on.

In the past 20 years, globalization has had a major impact upon science and high-
tech. In turn, the wide-ranging technological advances have changed the world and
affected the practice of management. The implications of this development are likely
to become more significant in the immediate future and present a new challenge to
scientists and engineers (Chap. 14).

We conclude with an introduction to the most relevant behavioral and management science concepts (Chap. 15) and various management training opportunities

(Chap. 16) that have proven particularly useful in nurturing an environment in which creative people and creative effort can flourish. Chapter 17 points to worthwhile reading matter, related to some of the major issues raised here.

2.1   A Word About Getting the Most Out of This Book

In the course of lectures covering the subject of this book, I have found it useful to ask frequently:

"Is this point relevant to some particular situation or issue which you have encountered?"

Whether the answer is yes or no, it is a most effective device to motivate reflection upon the matter at hand and thus relate it to your own experience. If the point in question does not seem relevant, it is useful to consider:

"Is it because you have not encountered this particular experience or that you have not thought about it before?"

The specific examples cited should bring to mind other examples and counter-examples from the reader's own experience. Such reflection has been responsible for some of the most valuable discussion. I am deeply indebted to the participants.

Insofar as this book aims to address creative scientists and engineers, and those managing them, we have chosen to address various practices and procedures such as recruiting or performance evaluation, by way of example. The particular examples were chosen for their expository value and should not be thought of as a recommended implementation. Rather, they should serve as a point of reference in understanding the procedures in place or under consideration.

3. Is the Management of Creative People Desirable?

This question will be explored from the point of view of the creative person considering management responsibility, and from that of the experienced management professional.

Most classical texts in management science would not admit the question about the desirability of managing creative people. What is more, people with extensive Hi-Tech management experience are likely to assert that any activity which requires the coordinated efforts of more than two or three people can benefit from good management.

3.1 Does Management Understand Creative People?

To explore this question from the viewpoint of someone who does not have this experience, one may postulate:

Management does not understand how to deal with creative people.

Early in their career, the majority of creative people would very likely agree with this postulate, since it is frequently borne out by their experience. This experience is most likely based upon having creative people, like themselves, as managers; creative people, who have as yet not learned to focus their attention upon the needs of the people whom they are asked to manage; creative people, who are still in the mode of narrow focus upon technical objectives.

Such inexperienced managers see their role as that of technical decision makers. This is not surprising since they were selected for demonstrated, superior technical judgment—not for proven management skills. After all, what can be the basis for judging a technical person who has never managed before?

Generally, such a negative attitude toward management will prevail until a sufficient number of individuals in a given organization have gained enough experience to benefit from a trial-and-error approach to management.

R. Kay, *Managing Creativity in Science and Hi-Tech*,
DOI 10.1007/978-3-642-24635-7_3, © Springer-Verlag Berlin Heidelberg 2012

Experienced management would advocate the subordination of the creative urges of the manager to those of the people to be managed. This view of management is not easy to implement. In fact, it is common practice to select managers in Hi-Tech organizations for their technical competence (often equated with creativity). That is to say, trial and error would have to be extended to the criteria for selecting managers. Most often, there is no basis other than technical competence as a selection criterion.

There is further reason for an essentially negative view of management on the part of creative people. Have not most of them been exposed to the view that "this is a university not a business," a view held by faculty who at the same time complain how "badly managed" their institution is.

The problems created by this attitude toward management can be alleviated only by increasing everyone's awareness of what management is all about. Let all scientific/technical people acquire the means of assessing their own potential and inclination to manage. They may thus hope to prepare themselves for the decision of whether or not to become managers when the time comes. They will then have the choice of being part of the problem or part of the solution.

It is our premise that such awareness will lead to the conviction that good management enhances the efforts of creative people, and that lack of management or poor management is a hindrance.

3.2 Creative People Do Not Need Management!

A second postulate based upon the views of more than a few creative people is:

Creative people are unique in their ability to achieve anything, including collective objectives. Therefore, there is no need for management.

It was implied above that the need for management is limited to the coordinated efforts of more than two or three people. But then, what do we mean by *managing* our own time?

Experience tells us that if we *manage* our own time, our only absolutely finite resource, we are likely to accomplish more than if we do not. This suggests that the concept of management is acceptable when we manage ourselves. But we question the concept when we are asked *to manage someone else* or are managed *by someone else*.

The most significant accomplishments of creative people are often the result of the single-minded pursuit of a particular objective. It is not surprising to find that in such pursuit, they are not likely to concern themselves with issues which are not motivated by the objective at hand.

From this, one may conclude that successful, creative people are unique in their ability to achieve outstanding results as long as they do not have to assume responsibility for others. Once they are dependent upon the cooperation of others, their ability to achieve results can be frustrated by potential conflict between group and individual objectives.

The opportunity of learning to develop the ability to resolve such conflict is often delayed due to additional years of education. During the 20 or more years of proving themselves in a very competitive academic environment, individual achievement has been the prevailing basis for success. The notion of thinking about how other people accomplish their objectives is not the dominant mode of thinking about problems.

To illustrate this point, let us see how one creative scientist arrived at a first awareness of management responsibility.

Frank is now a successful manager in one of the most respected industrial research laboratories. Before that, as a well-established chemist, working for some years as an independent researcher, he had always been fortunate enough to have a technician to help him. For many years, he thought of these technicians in terms of their ability to help him accomplish his objectives. Some were better at this than others.

Not until *several years* after he had left the academic world was he exposed to some form of "management training" in an organization which placed value upon such training. He was still operating in the mode of an individual, independent researcher, pursuing his own research interests with the help of a technician.

Thinking of himself as a person with, if anything, an *above-average* sense of social responsibility, he was shocked to find that the idea of responsibility for his technician had *never* crossed his mind. He had clear ideas about the technician's ability to help him accomplish his own objectives. On the other hand, Frank had *never* given thought to his own role in helping the technician accomplish *his* objectives. In fact, they both reported to the same manager, who thus took responsibility for the technician. The chemist's role had been to tell the manager how well the technician satisfied his needs.

By implication, we may have suggested that creative people have problems in managing others, but have no difficulty in managing themselves. Experience suggests otherwise. Some outstandingly creative people are "driven" to succeed, work excessively long hours, and neglect every other aspect of life. Some make unreasonable demands upon their colleagues to make up for their own inadequacies. Not that these traits are the sole province of creative people. On the other hand, it seems fair to say that the unique talents of outstandingly creative people do not necessarily extend to managing their own affairs—or those of others.

To summarize: The first contact creative people have with management often reinforces the preconception that management does not know how to deal with creative people.

There is no evidence that creative people have a unique talent, for managing cooperative effort or for managing their own affairs. Since they are uniquely creative, a trait highly valued by society, they are more likely to get away with a lack of management concern, by having others compensate for their shortcomings.

3.3   Characteristics of a Creative Hi-Tech Professional

Let us look at a successful, creative Hi-Tech professional to understand the basis for some of the assertions just made.

3.3.1   Recognition as a Prime Motivator

To achieve success in a highly competitive profession, one is forced to specialize. The narrower the field of specialization, the greater the opportunity to get to the top. This may be the top of the immediate group, the department, the organization, or ultimately the top of the profession. Successful specialists join professional societies which have a section devoted to their subfield. To win recognition in one's chosen field is a major motivator of many successful, creative people.

At the extreme, it serves to satisfy inflated egos and drives people to narrow their interests in order to become Number One, if only in a narrow subfield. Preoccupation with the desire to win recognition can go so far as to make people suspicious of their colleagues, based on the fear that they may appropriate each other's ideas.

3.3.2   Achievement for Its Own Sake

A characteristic of many creative people in Hi-Tech endeavors is that they derive great satisfaction from having solved, or even better, having defined and solved, a difficult technical or scientific problem. Anything which limits the perceived level of independence required to define and/or solve the problem at hand is considered an obstacle to success and a potential demotivator.

Outstanding achievement implies high standards of excellence. An exaggerated standard of excellence, "perfectionism," is not uncommon among high achievers. When carried to the extreme, perfectionism can be a guaranteed road to failure. It is always possible to set a goal which is beyond reach and thus makes failure unavoidable. While not a common trait, perfectionism is likely to be found more often among potentially creative professionals than among other groups.

When high achievement is the motivator, it is incumbent upon the problem solver to work efficiently. He or she is apt to be in competition with the best for early publication, invention disclosure, or market readiness. This need for efficiency can at times detract attention from effectiveness. Doing things "the right way" often overshadows concern with doing the right thing. The elegance of the solution may become more important than the significance of the problem.

To ensure future success, successful specialists must increase their knowledge. This must be done in a systematic and efficient way, or the successful problem solver becomes one who studies problems, an activity which generally wins little recognition. Personal interactions often are designed to enhance problem-related knowledge. It is the rule rather than the exception among Hi-Tech professionals to talk shop while eating lunch. Others may talk about sports, cars, hobbies, or family-related subjects; the specialist seems more comfortable to stick to work-related topics. "Relating" to one's colleagues is generally confined to work-related subjects.

3.3.3 Achievement Versus Relationships

In the academic environment, personal relationships are further burdened by the need for frequent review and evaluation of the work of one's colleagues and peers. Discussion of qualifications for an appointment may be influenced by opinions based upon personal relationships, but these opinions are rarely voiced. Such decisions are held to be the proper province of judgment limited to professional competence, an area in which each participant feels uniquely qualified by virtue of membership in the deciding body.

The above characteristics tend to promote a solitary work-style and place little emphasis on developing personal relationships beyond what is necessary to get the problem solved. It is not unusual for Hi-Tech professionals to see each other every day for a period of 10 or more years without knowing anything about each other's personal life. I have seen long-standing colleagues turn their back at the first sign of a personal problem. It is probably fair to say that the predominant relationships between Hi-Tech professionals tend to focus upon their common technical or scientific interests.

Compare this with the motivation and interaction of the homemaker who seeks interpersonal relationships with the parents of their children's schoolmates to find out with whom the kids associate. Not provided with a ready-made collection of people who share common professional interests, the homemaker is forced to develop interpersonal relationships in order to find people with common interests. In the world of the homemaker, people are less constrained in their choice of people with whom they associate; hence, interpersonal relationships become as much a factor as common interest.

Recognizing these characteristics can help us understand the issues faced by the successful creative professional when he or she first assumes management responsibility. Often, the characteristics that have made this individual successful are so deeply ingrained that the new manager attempts to continue his/her technical activities by working excessive hours with the likely result of early burn-out or unsatisfactory performance in one or both areas. The fear of technical obsolescence adds an additional impediment to the willingness of the creative individual to make the change from solving technical problems to creating an environment in which others can achieve. Without training or help from an experienced manager, it is very difficult indeed to make the switch from doing the work yourself to the job of assessing the match between the job to be done and the people available to do it, and evaluating whether a given task can be done with the technology and resources at hand.

Thus, the initial question may be rephrased to read not whether management of creative people is desirable, but how much management and by whom. The answer that we offer at this point: the less the better and by the most qualified people available.

This answer is based upon a view of management, particularly as it relates to creative people, that will be described in the next section.

3.4   Managing Creative People: Our Hypothesis

To maximize creative potential is the concern of each creative individual. It is the particular concern of those responsible for the management of efforts involving creative people.

Those who feel motivated and qualified to do creative work consider their ability to influence, if not control, the direction of their work as their most valued privilege, the most important factor in developing their full potential.

Managers of Hi-Tech enterprise must protect this privilege while simultaneously meeting their prime responsibility, which is the welfare of the organization which makes creative effort possible. This may be the firm served by the project, department, or laboratory, or the university or government agency which assumes responsibility for a particular project or laboratory.

If the mutual benefit to the organization and to the creative individual is to be maximized, individuals must be involved beyond the mere pursuit of their creative technical work. The objective of this involvement should be the convergence of the interests of the individual and the organization. An important duty of management is to facilitate this involvement (Leonard 1999).

Some of the most successful creative people have chosen deliberately a role which minimizes their management responsibility in order to concentrate on their own work. On the other hand, the success of the organization depends, maybe to an even greater extent, upon technically competent people who possess a relatively rare human quality, that is, *the ability to stimulate technical accomplishments which benefit and earn recognition for others.*

3.5   Creative Team Effort

The following account provides an example of a successful creative team effort which should serve to elaborate our hypothesis.

Len was brought into the research laboratory from another location of the organization, to take over a group of 20 people. For more than 5 years, this group had had little impact upon the organization or their field of science. We here define "impact of research" in terms of influence upon the direction of research in the field, and/or research leading to an important application.

While all of the members of this group were considered competent and motivated, their individual efforts had no cohesion or common goal. As in so many areas of high technology, in the field in question, significant ideas often need to be evaluated in a complex problem space and require considerable follow-up effort to demonstrate feasibility and eventual success.

Len insisted upon a project proposal which would focus the efforts of the group upon a common goal. By implication, this single goal would be the integral of the collective "best" ideas of the group members. It made it necessary for people who had spurned team effort to begin to share knowledge and ideas in order to come up with an acceptable proposal.

Len made himself unpopular by insisting for many months that he would settle for nothing less, and, as a last resort, would consider defining a project himself. During this period, Len encouraged every idea involving more than a single person and thus got a feel for what was promising.

Morale dropped precipitously, one member of the group resigned. But, after 7 months, a proposal emerged which had the ingredients of potential success. The goal had real significance to the organization and, if successful, would open a new area of research. The timing was opportune; more traditional approaches had encountered fundamental limitations.

Len's persistence paid off dramatically. It ultimately led to strong convictions on the part of the project members about the goals of *their* project. Over the next 5 years, the project grew significantly. Many members of the group gained stature in the profession by having created a new field of research, pursued in every part of the world. New products emerged from this effort and some of the group members availed themselves of opportunities to assume responsible positions outside the research organization.

Len did not tell them what to do, but urged them to consider the potential value of concerted action. He understood that creative people derive their principle satisfaction from doing creative work and getting recognition for it. They are not *motivated* by the lack of external pressure or the ability to work in solitude. External pressure and collective action, which may necessitate formalized procedures, are only rejected insofar as they *interfere with* creative work.

Many, not all, creative people who are confident of their own abilities thrive under pressure—they welcome a challenge. People who have reason to lack confidence in their creative ability often focus upon external influences, such as pressure or lack of individual freedom, as a way of rationalizing their lack of effectiveness.

3.6 Individual Creative Effort

The above example of outstandingly successful creative team effort is not intended as an argument against individual creative effort. It is meant only to support the contention that in the world of Hi-Tech, creative team effort plays a very important role.

As it turned out, Len had to deal with two creative individuals, so considered by virtue of their past contributions, who could not be integrated into this team effort.

One of the members of the group who had not been able to accept the selected approach, which was orthogonal to his own ideas, was offered a sabbatical which she accepted. The other individual, whose past work had great influence upon the proposal, was neither acceptable as a manager nor able to accept any of the managers appointed by Len. He played an effective role as staff advisor to Len which allowed him to continue to pursue his individual efforts.

> The rarest and most valuable quality of solitary creative effort is the dedication and sustained motivation required to pursue goals which involve high risk over long periods of time. Such dedication is driven by conviction strong enough to compensate for the lack of reassurance inherent in the shared objectives of the team.

One of the most difficult management tasks in the world of Hi-Tech is to understand the motivation for the pursuit of solitary goals: Is it commitment to a worthwhile idea, or is it a manifestation of egocentricity?

Generally, past performance is the only guide. Major research universities provide the best opportunity for the unproven scientist or engineer to pursue such solitary objectives. But every professor knows that only a good batting average will assure continued access to uncommitted research funds.

This view of the subject of Management of Creative Effort has been presented at this point to prompt you, the reader, to think about your attitudes toward the subject of management and your role as a manager before going on with the book.

3.7 Summary

We have recognized the basis for some negative attitudes toward management on the part of creative professionals. Both the university experience and early encounters with management tend to shape these attitudes.

Some of the personal characteristics of creative individuals tend to promote solitary pursuits rather than team effort. The value of both individual and team effort in the pursuit of creative work is recognized.

Our hypothesis: Creative effort requires and can be facilitated by what we define as effective management. A better understanding of such management practice on the part of the creative individual is a worthwhile aim.

The questionnaire is designed to help you assess your motivation to assume a management role. Review of the filled out form after reading the remainder of the book should reveal to what extent you have had occasion to change your views.

Questionnaire: To Be or Not to Be a Manager

My Reasons for Wanting to Become a Manager

Score					1 = Does not apply to me at all, 5 = Most relevant
1	2	3	4	5	Advance Career; Get to Top
1	2	3	4	5	Hedge Against Failing Creativity
1	2	3	4	5	Increase Options in Future
1	2	3	4	5	Potential for More Satisfying Career
1	2	3	4	5	Increase Control Over My Own Work
1	2	3	4	5	Get People to Help Implement My Ideas
1	2	3	4	5	Enhance Ultimate Earning Power
1	2	3	4	5	Improve Working Life for People Like Me
1	2	3	4	5	Pressure from Family for Status, Money
1	2	3	4	5	Live Up to Expectations of Parents

My Reasons for NOT Wanting to Become a Manager

1	2	3	4	5	Have No Talent, Wrong Personality
1	2	3	4	5	Do Not Want to Dilute Creative Efforts
1	2	3	4	5	Do Not Want Responsibility for Others
1	2	3	4	5	Want to Develop Technical Competence
1	2	3	4	5	Not Ready to Tell Others What to Do
1	2	3	4	5	Do Not Want to Do Something I Don't Believe In
1	2	3	4	5	Family Pressure to Pursue Nonmanagement Career
1	2	3	4	5	Want to See How Far I Can Go on Technical Ability
1	2	3	4	5	Hate Administrative and Paper Work

4. Managing Your Own Work

This topic is motivated by two objectives: First, to identify some generally applicable management issues, of interest to both managers and non-managers. Second, to provide an opportunity to reflect upon our "intuitive" management style.

4.1 Objectives, Activities, and Results

4.1.1 The Basis for Decision Making

To undertake a new task or to terminate or set aside an ongoing task involves a decision. The decision may involve a personal choice among several possible tasks or priorities dictated by others.

However superficial and unrefined the analysis which precedes the decision, we have some understanding of the objective of each task we are to undertake. Already in graduate school, we have encountered a wide range of objectives.

- Get best possible grade on assignment due next week
- Get passing grade on next exam
- Prepare for decision on thesis project
- Get maximum feedback from thesis presentation
- Meet commitment to completion of assigned task
- Help students in section pass their exams

It is important to distinguish between the *activity* of "working on an assignment" and the *result* expected from this activity, "obtaining a grade." The objective relates to the result.

- Researching a thesis project vs. preparing for a decision
- Making a thesis presentation vs. getting feedback
- Working on an assignment vs. meeting commitment
- Teach a section vs. assuring that students pass exams

R. Kay, *Managing Creativity in Science and Hi-Tech*,
DOI 10.1007/978-3-642-24635-7_4, © Springer-Verlag Berlin Heidelberg 2012

If we have no concrete concept of the objective, we take the risk that the *activity* may not lead to the desired *result*.

For example, in your thesis presentation you are more likely to elicit feedback if you indicate some of the alternatives you have considered and the reason for discarding them, or suggest some of the unanswered questions. Or in the case of an assigned programming task—should you design a given program module based upon a new idea you have had, or on the basis of a proven algorithm? The former is likely to be more fun and produce spectacular results; it may not work at all and/or may take more time than you can afford. The latter is sure to work and the time of completion fairly certain. What are the desired results? Something spectacular with a risk of failure, or the timely delivery of a working module? In the case of thesis selection, your objective is to optimize your choice in terms of the best possible result, based on available information, resources, and time.

What are some of the criteria for having met your objective of selecting a thesis topic which will yield the best possible result?

- The idea appears to be original
- If successful, this idea will have a significant impact, e.g., others will drop what they are doing and jump on this idea
- This idea will provide an opportunity to demonstrate creativity with a minimum expenditure of time
- It is within your personal ability to do enough to demonstrate the value of this idea to others
- It is likely that the work required for initial demonstration of feasibility is within practical bounds

Not every decision we are called upon to make warrants such detailed analysis. But, as in the case of study habits, the earlier in our career we can develop some proficiency in the process of decision making, the greater the potential benefit.

4.1.2   Our Intuitive Management Style

Analysis of our decision-making process will not only serve to improve our ability to make decisions, but it will provide some insight into our "intuitive" management style:

- Do you naturally seek the advice of others or do you prefer to think for yourself?
- Are you more attracted to a project because you can do it alone?
- Are you realistic in estimating your own ability and the time required to accomplish some tasks?
- How do you approach this matter of estimating?
- What is the value system at work when you set objectives for yourself?
- Among the many promising problem-solvers there will be only a few promising problem-definers. In these terms, how do you think of yourself?

In this very competitive Hi-Tech arena, most "creative" approaches will involve some element of risk. It is important to find out what level of risk one is comfortable with. The opportunities for discovering this are none too frequent. Hypothetical situations lack the decisive element, namely the emotional aspect of risk taking. Thus, it is well to avail oneself of every opportunity to make or take part in risky decisions.

Having come to recognize one's intuitive management style, one is better able to take advantage of strengths and learn to compensate for weaknesses.

4.2 Efficiency and Effectiveness

Another useful distinction to draw involves the concepts of efficiency and effectiveness. Efficiency relates to the *activity* we are engaged in. Effectiveness relates to the *results* we are trying to achieve.

We can be very efficient in preparing for an exam or finishing an assignment, but totally ineffective by having misunderstood or neglected to ascertain the extent of the subject to be covered.

It is most likely that in graduate school our thesis advisor proposes the topic or project we are involved with. Early in our postgraduate career, we may still count upon someone else to take the ultimate responsibility for our project. But at some time, it is necessary to assume the responsibility for defining the objectives of a project, in order to decide the worthiness of its pursuit.

More mature graduate students proposing a thesis topic will show concern about the objectives, potential impact, and the amount of help required. One of the more valued recommendations from a professor points to the significant role played by the student in the selection of the thesis topic.

The seasoned manager in evaluating an applicant's resume looks for achievements, not job descriptions.

The maturity which you project when seeking the position that is to launch you upon your chosen career will reflect the thought you have given to such issues (see also Sect. 8.1 on Recruiting).

4.3 Dealing with the "Incompetent" Project Leader

One of the earliest opportunities of dealing with management issues presents itself in our interactions with a negatively perceived manager, supervisor, professor, or project leader.

Pursuit of this topic could well take us to every aspect of management. Let us here be content to note that an attempt at objectivity and an analysis of the part we ourselves are playing in this interaction can be very rewarding. It can be a most valuable learning experience involving our own reactions and limitations. It can also provide some of the immediate gratification of having solved a sticky problem.

Example: You have proposed a novel approach to the leader of your project. The response: "I'll think it over." Every subsequent try to get a definitive reaction fails. You are stymied, since you need the cooperation of others to proceed. You may seek understanding among your colleagues. Some will and some will not encourage you in the conviction that the particular project leader is, among other things, unreasonable and incompetent. Your natural tendency is to listen to the colleague(s) who agree with you. The fact of the matter is the people who disagree with you are more likely to provide you with some understanding of the project leader's point of view. If your proposal is not well received by your colleagues, the project leader may feel it is better you are dissuaded by your peers.

Convinced of the technical merits of your proposal, you may ask yourself whether your manner of presenting the matter may be at fault. Does the relative merit of your proposal depend upon pointing out the inadequacy, inferiority, or stupidity of the previously accepted approach? Who was responsible for the previous approach? Who did accept it? The answers to these questions may be decisive in resolving the matter. Should you have sought these answers before you proposed your solution? How far are you prepared to go to affect a resolution? Is this the most important issue which you must resolve at this time vis-à-vis your colleagues, your project leaders? Are you using up precious goodwill? Is it worth it?

Some few people have a natural talent for dealing with such issues. Many of us don't. Most of us could improve our ability considerably. We may do this by trying to be objective, by not pushing these problems aside or into our subconscious where they are likely to become aggravated. Some people are fortunate in being able to develop personal relationships with their colleagues who provide them with solicited, or even better, unsolicited feedback.

The "experience" which we are generally supposed to lack may be due to our being unaware of the learning opportunities around us. It is probably easier to learn from the weaknesses (recognized as mistakes) of others than from their strengths. The mistakes are more obvious and the strengths of our peers may in any event be beyond our reach.

4.4 Work Habits of the Successful

We can learn work habits from those who have experienced success. Most scientists and engineers who have met with consistent success over a number of years are very disciplined in the way they manage their work. They have experimented enough to find the approach they are most comfortable with.

Typically, they will divide their time between a variety of tasks. They are likely to spread their risk. They will work on a project with a high probability of success as their main task, and devote some time to the exploration of a more risky but potentially more significant idea. Besides doing research, they will spend time reviewing for some journal, an excellent way to become part of the prepublication community and of keeping up-to-date in one's field.

An experienced development or manufacturing engineer will commit a portion of his/her time to maintaining skills not frequently used and to developing new ones, as well as to keeping up with the field or learning something about a related field.

Forming a habit of working in a "multiprocessing" mode has several potential benefits. It provides a change of pace when one has reached the point when progress seems to have slowed down. It also affords the opportunity to evaluate alternatives, be it alternatives to a particular problem, alternatives to working habits, or alternative lifestyles.

While I do not believe that there is a generally applicable method of managing one's work, more conscious concern with the issue seems to payoff. See also: Drucker (2007).

4.5 Publications

Professional reputation is often based upon material published in the open literature. One widely adopted measure of the impact of published work is the number of citations of a certain publication over a given period of time. Professionals whose output is reflected in the papers they have published in the open literature are increasingly being evaluated in terms of the impact of their publications.

Considerable research has been done, based on the Citation Index, as to the most frequently cited publications in nearly every field. Libraries also use this information as a guide in deciding upon subscriptions. This, of course, is a consequent factor in the distribution of citations among publications.

There are other considerations which influence the choice of journal to which a prospective publication is to be submitted. In the case of very timely research results, the average delay between first submission and ultimate publication date is an important factor.

Experienced professionals, and university faculty in particular, are aware of the journals which facilitate maximum impact and/or early publication in a given field. Academics, more than any other group, are dependent upon publications as a means of achieving effectiveness.

The activity of publishing does not guarantee the desired result of maximum impact or timely recognition; choosing the right publication helps.

4.6 Managing the Use of Time

Earlier on, the point was made that time is our only absolutely finite resource. Most resources available to us can be increased or augmented. We can have more or better machines to help us; we can increase or enhance our access to knowledge; we might even be able to enhance our physical or mental abilities. But the use of our time can be enhanced only through good management.

Managers occasionally complain that they have little control over their time. Real-time events often take priority over planned activities. Some of these same managers have difficulty scheduling their time when on a vacation. In fact, they do not enjoy vacations because they have to structure their own time. "Event-driven" managers may project the image of the ultimate in responsiveness; it might also be interpreted as an indication of their reluctance to assume responsibility for managing their own time.

A first step toward the improvement of your skills in managing your time is to develop insight into your own habits. Short daily entries in a diary, which describe the nature of each separate activity, are most revealing. Some creative people maintain such diaries throughout their career and use them for periodic review (weekly, monthly) of their activities.

Your need or dislike of structure in managing your time provides another useful insight. Accordingly, you may have to pay appropriate attention to insure convergence of your interests with the interests of those around you. As a manager, this insight is not only useful, it is essential.

4.7 Supporting Decisions

To the extent that our activities involve other people, we may encounter the need to support decisions that conflict with our own position. In this context, "other people" refers to those dependent upon us or upon whom we ourselves depend.

Let us try to understand the variety of situations which give rise to this problem by some examples.

We are mountain climbing with a group of people and must choose between alternative routes. We probably have little problem accepting a choice other than our own:

- If the group has done a lot of climbing together, there has evolved a generally accepted procedure for making such decisions. If the procedure has been followed, we are likely to accept the decision.
- The group has not climbed together before; it is not clear that there is a generally acceptable way to make such decisions. Should we establish some rules before we leave?
- The group decision has followed the established procedure, but my personal interests are at stake. I had made commitments to return by a certain time, which cannot be met accepting the group's decision.
- Everyone has made a proposal for the next climb by the group. My proposal was not chosen. Am I going on that climb?
- A decision has been reached that I consider to endanger the safety of the lead climber. Do I abide by the decision if I am the lead climber? If I am not the lead climber?

However the decision has been arrived at, and whoever was previously for or against, it is absolutely essential that every member of the group accepts and fully supports the decision once it has been made. The only alternative is to forgo participation.

The person who assumes responsibility for the group's activities cannot expect everyone to agree with a given decision. But the group leader has every right to expect full support of the decision by those who want to participate in the group's activities.

Problems are usually encountered because of our inability

- To make unpleasant decisions
- To make decisions which are at odds with those of our peers
- To accept decisions which are at odds with our interests
- To express ourselves forcefully or to carry out a decision we did not support

Such problems are fairly common among talented professionals. They limit the effectiveness of the individual and of the group.

We shall return to the subject of decision making, subsequently. It is one of the most important management responsibilities. A manager who has difficulty making decisions will fail, in spite of excellent marks in every other subject.

4.8 Summary

We have identified the need for clear objectives and the need for differentiating between activities and results as generally useful concepts in making decisions about our own work.

The example of the negatively perceived manager suggested that, before we are able to effectively change a situation, we must first understand how it came about. Without feedback from others including those who disagree with us, it may require painstaking analysis.

We can learn from the experience of successful practitioners of our art that working in a multiprogrammed mode has certain benefits.

If publication of our work is significant to our professional career, attention should be paid to where we publish. The management of time is probably the most important factor contributing to our effectiveness.

We recognize the importance of the process of making decisions and the problems we face in supporting a decision in conflict with our self-interest.

All of the above should prove useful to us individuals—for managers they are essential.

5. Desired Qualifications of Managers

How do experienced managers select new managers? Are there general criteria for the selection of managers? In this chapter, we shall look at criteria for selecting people for entry-, mid-, and top-level management.

5.1 Picking the "Best" Available

Jim, my manager, comes to my office first thing Monday morning. Without the usual inquiry about the family or the weekend happenings, he gets to the point:

"The group is getting too big for me to manage by myself. I have been asked to take on additional responsibilities. We will organize the group into two separate projects and I would like you to become manager of project A. How about it?"

What do I say?—"Why me? I am having a good time doing my thing in the lab.— Management is not for me, at least at this point.—Paperwork is something foreign to me.—You mean all this work in graduate school down the drain?—Do we really need *more* managers?"

Jim listens patiently, to what he later termed "the expected response." He explains in more detail the need for this additional level of management. The conversation ends with him asking me to think this over; there is a casual reference to my colleague, Robert, who would be the alternative project manager, if I turned down the job.

It is immediately clear that while neither Robert nor I are the most obvious management candidates in the organization, of the available people, I am certainly the best qualified person to manage project A. What is more, I can't see Robert in this job at all, especially my reporting to him.

The first lesson this new-manager-to-be has learned about management: you deal with the reality of the situation bounded by the constraints of time and available resources.

Subsequently, Jim told me that he had given the matter considerable thought over a period of time. He had in fact tested my reaction to various situations in the group

R. Kay, *Managing Creativity in Science and Hi-Tech*,
DOI 10.1007/978-3-642-24635-7_5, © Springer-Verlag Berlin Heidelberg 2012

in the past. He saw it as one of his most important tasks as manager to assess the potential of the people reporting to him and he felt confident that I could handle this assignment.

Are we avoiding the issue of management selections?

No.—We only want to make a point: An important aspect of management selection is the personal knowledge of potential candidates by the person whose own success depends upon selecting the best person available.

> Very often, what appears to be nepotism is simply a manifestation of the natural instinct to seek those upon whom we most depend, from among those we know the best.

By implication, if you want to be considered for a management position, you must be visible to those responsible for selection.

Some large organizations have found ways of optimizing the management selection process on a more global scale. They go to considerable lengths to identify the most promising people and, early in their career with the particular organization, bring them to the attention of high-level executives.

Smaller organizations often are dependent upon outside recruiting for management positions. They may in fact use such positions to attract people who can bring new skills into the organization. The most common problem encountered is due to the fact that the person under consideration is evaluated on the basis of past performance in another environment, rather than potential to do well in the new environment.

Inexperienced management frequently does not take the task of recruiting seriously enough. At the very least, people at all levels of the organization should contribute to the decision to be made. This may be burdensome, but failure to do so denies the organization the shared learning experience, which will be the basis for better decisions in the future. What then, are some of the desired qualifications at the various levels of management in Hi-Tech organizations?

5.2 Qualifications of Project Manager

At the first or lowest level of management (sometimes referred to as the entry level), technical competence, or better, technical excellence is the prime requirement. This manager is the person who makes day-to-day decisions about technical alternatives based on his/her knowledge of the subject. When the superior knowledge of the project manager is recognized by members of the project, decisions are more readily accepted.

Dependence upon the technical competence of the project manager can be frustrated by the lack of rapport between the manager and the members of the project. Ideally, Hi-Tech managers at any level are able to use their technical insight to encourage the best ideas advanced by members of the project, department, or organizational unit. Negatively perceived personal relationships between the manager and subordinates, such as lack of respect, reduce the acceptability of the manager's viewpoint.

Above all, managers should not take advantage of their position to preempt project members from presenting their ideas first. Nor should they give way to the temptation of regular second guessing of ideas in their incubation stage.

A truly successful manager will always try to recruit the most competent people to be found, with the expectation of finding some more competent than himself or herself. A manager, who has developed the reputation among the members of the project of championing *their ideas*, will have no problem in dealing with the outstanding individual.

Entry-level managers trained in science or engineering generally bring little in the way of administrative skills to the job. Examples of necessary administrative skills are organizing your daily work, evaluating the performance of others, dealing with requests for information from other parts of the organization, maintaining schedules, conducting meetings, preparing a budget or proposal, making presentations, and recognizing the essentials in the presentations of others.

The most common complaint among entry-level managers relates to what is perceived as excessive amounts of administrative work. Few of us are born with administrative skills. They can, however, be acquired. The administrative burden will decrease with the level of skill we can bring to bear. Administrative skills are in some ways analogous to study skills; the sooner they are acquired the more benefit can be expected.

While the first-time manager is not expected to possess these skills, an understanding for—and readiness to develop—such skills is expected.

Dealing with people is likely to present the first-time manager with the greatest challenge. Should she or he find it unduly difficult to deal with this aspect of the job, return to purely technical work is easiest at this level. In accepting a first management job, one should always keep this option open. While experienced managers are expected to have developed the ability to recognize management potential or the lack thereof, in the case of a first-time manager the judgment involves a great deal of subjectivity. Particularly among creative people assuming management responsibility for the first time, I have encountered such negative reaction to the need to relegate their own work to second place that to some, management responsibility can become an unbearable burden.

The technical competence of a professional is most often associated with superior accomplishment in a given field of science or technology. Barring overt personality problems, technical competence therefore plays the dominant role in the selection of first-time managers.

There is the widespread view, rarely challenged, that the best salesperson invariably makes the best manager. This view gains credence from the fact that one of the essential talents of a successful salesperson is the ability to deal with people. On the other hand, it is possible to be an outstanding scientist or engineer, devoid of this talent.

There have been Nobel Prize winning scientists who have failed completely as managers. Furthermore, it is generally easier to define and quantify the objectives of a salesperson, and hence measure relative performance, than to compare the relative accomplishments of two engineers working side by side, one responsible for design, and the other for implementation, or of two scientists, one doing analytical, and the other, experimental work.[1]

We here have drawn attention to the problem of selecting managers from a group of scientists or engineers who by definition have had no management experience.

The problems encountered by the first-time manager (with suggestions of how to deal with them) will be addressed when dealing with project management in the next chapter.

5.3 Qualifications of Department Manager

At the middle management level, where the manager has responsibility for several entry-level managers, there are requirements beyond that of success at the entry level. Here the ability to motivate people and to understand the potential impact of the various projects upon the total organization is the prime requirement.

This individual has already proven him or herself in an entry level management position and thus has met the criterion of technical excellence in some specific area. It is not reasonable to expect a person to be equally qualified in other technical areas. Hence, the middle-level manager must develop the ability to depend upon others for specific insight and judgment. It is a common failing of newly appointed second-level managers to feel the need to know at least as much as everyone else on every subject of concern. Since this is usually not possible, it is better to develop the ability to judge the dependability of the people available for advice.

At this and higher levels, it is often necessary to reallocate resources between projects. Since this involves taking away as well as giving, a middle-level manager must, of necessity, be prepared to reduce resources without crippling a project. This turns out to be a much more difficult task than starting a new project. Increasing resources tends to increase the knowledge which provides the basis for future decisions. Shifting resources from a lower priority project is likely to further decrease its relative success potential. Hence, it may be preferable to terminate such a project.

The middle-level manager is also expected to take on specific tasks on behalf of the next-higher-level manager. These may relate to one of the special areas of skills or interest represented by the middle-level manager's department. The middle-level manager must also be able to identify issues and opportunities which are beyond the concerns of daily operations or current objectives. These abilities of the mid-level manager are also called upon in connection with the developing of a strategy and operating plan (see also Managing a Department, Chap. 7).

---

[1]It is interesting to note that "people-skills," on the part of aspiring physicians, have not been part of the selection process until very recently (Harris 2011).

Above all, the middle-level manager becomes the example and mentor for the newly appointed entry-level managers. His or her ability to identify and develop management potential is crucial to an evolving organization.

5.4   Qualifications of the Top-Level Manager

At the highest level, the manager's task becomes considerably broader. Not only is the top executive expected to have performed well at the lower levels and hence appreciate the relevant issues, but the person at the top also sets the "tone" for the organization or organizational entity for which he or she is responsible.

Setting the tone is a difficult thing to define in concrete terms. Anyone who has observed diverse Hi-Tech organizations is quickly aware of the prevailing *Tone*. The manner in which meetings are conducted, the accessibility of top personnel, the appearance of the physical facility, the manner of dress, punctuality, deference to rank or the lack thereof, respect for the individual, etc. To the extent that the top manager provides the role model, he or she will set the tone.

Lack of conscious tone-setting is generally felt as lack of identity, lack of predictability, and lack of stability, i.e., a generally negative perception. Organizations which have succeeded in the *search for excellence* are invariably identified with a positive image which emanates from those who set the tone.

Setting the tone is sometimes thought of as defining and molding the "culture" of the organization, an equally fuzzy concept. A working definition of organizational culture would include a set of beliefs, policies, and guiding principles (written or otherwise). These principles are most important in a rapidly evolving and changing environment as commonly encountered in the Hi-Tech world. They allow for consistent decisions and conflict resolution in the absence of explicit rules and regulations. They allow for an *interpretation* of existing rules that are not explicit enough or appropriate to deal with new or especially complex situations. These cultural values evolve through common experience.

Since this is an aspect of management with far-reaching implications, for both the individual and the organization, we shall return to the subject of Organizational Culture in Chap. 13.

The top-level manager determines policy, defines strategy, and has a direct responsibility for the success of the enterprise. To do this, the top-level manager must have a thorough understanding of all of the internal and external factors which affect the total enterprise. This is a tall order in the world of Hi-Tech, since these factors have a habit of changing at an ever-increasing rate. It is therefore among the most important tasks, for the person at the top to develop people upon whom he or she can depend. The ability to learn and the broadest possible interests (as contrasted with highly specialized interests) are also among the qualities sought in a top-level manager.

The need to gain the complete confidence of those who have the ultimate responsibility for the welfare of an organization increases with the level of management.

Decisions made at the highest level of management must often be based on less information than one could wish for; this in spite of the fact that these decisions are usually the most far reaching. In making these decisions, the top-level manager depends upon others for information and opinions. Nothing influences the top-level managers more in their judgment of subordinates than the quality of information and opinions made available for their decision making.

Probably the skill most difficult for the Hi-Tech professional to develop, at every level of management, is the ability to delegate—to delegate not only responsibility, but authority to make decisions. The lack of confidence in others to do the job adequately, or as well as oneself, is the most common failure mode among Hi-Tech managers.

If the task to be delegated addresses a recurrent issue or is likely to require follow-up, the argument "I can do it faster myself than explain it to someone else" loses all validity.

Reluctance to delegate suggests inability or lack of readiness to assume broad responsibility which is unacceptable at the top of a Hi-Tech organization.

Another hallmark of a top-level manager is a readiness to accept honest mistakes on the part of subordinates. An effective manager turns mistakes into learning experiences and in the process strengthens personal relationships by a willingness to share responsibility for correcting the problem.

The concept of "dimensions of personality" (Levinson 2006) provides a useful aid to judging the personality traits of someone to be considered for the top management position. Levinson identifies 20 qualities including

- Tolerance for ambiguity; can stand confusion until things become clear
- Judgment; knows when to act
- Achievement; oriented toward organization's success rather than personal aggrandizement
- Maturity; has good relationship with figures of authority
- Integrity; has well-established value system which has been tested in various ways in the past

For each such quality, Levinson identifies five levels of valuation. For example, for the quality

- Adaptability; manages stress well. The five levels are:

  1. Doesn't tolerate stress well. Many personal, physical, and family symptoms.
  2. Can take stress well if supported by others. Worries a lot about what might happen.
  3. Does reasonably well under bursts of stress but not under long, sustained pressure. Worries in a healthy way about solutions to problems.
  4. Can take sustained pressure with normal symptoms. Has effective coping devices, such as consultation with others' and taking time off.
  5. Takes whatever comes in his or her stride and seems to thrive on it.

Levinson recommends caution in applying such an evaluation with the expectation of quantifiable results. He suggests that it may be more appropriate to consider "What profile of dimensions best fits the profile of behavior required for the job description I have drawn up?"

There are occasions when no appropriately qualified person is available to fill a management position. Experience suggests that it is preferable to have the next level of management assume the additional responsibility in an "acting" role, rather than install an inadequately qualified person on a temporary basis. The acting manager will be highly motivated to find a properly qualified person; the temporary manager may be more interested in proving him/herself.

To summarize the principal requirements of the various levels of management responsibility:

- Entry level: Technical competence relevant to the project at hand and interpersonal skills which make full use of the competence of the members of the project. Ability and desire to develop administrative skills.
- Middle level: Technical breadth, rather than depth. High interpersonal skills. Ability to motivate and influence people. Well-developed administrative skills.
- Top level: Ability to learn, in a variety of disciplines: technical, administrative, and political. Ability to delegate. Secure with a set of values which can serve to build the desired image of the organization. Ability to win the confidence of subordinates, peers, and superiors. Ability to make decisions in the face of uncertainty.

In subsequent sections, we shall look at some of the ways these skills and abilities can be developed.

5.5 Leadership: To Lead or to Mislead

Leadership is a quality which is desirable at all levels of management. It is essential at the top. Leadership is not bestowed upon an individual, it is assumed.

When we think of effective leaders, examples of those who have led and those who have misled come to mind. What makes them leaders? What differentiates them?

All leaders command a following. People trust them and are prepared to accept the beliefs of the leader. They are generally ready to do this without knowing the leader personally. Frequently, they are willing to accept a projected image of the leader, accurate or otherwise. Most people place considerable confidence in the leader's judgment and are prepared to make significant personal commitments, involving time, money, and even personal risk.

There can be no doubt that the followers identify with the objectives which the leader professes. Consciously or otherwise, people believe that the pursuit of the leader's goals will benefit them and help them realize their aspirations.

Peter Drucker (1988) has identified the factors which distinguish the leader from the misleader:

- The compromises that a leader makes with the constraints of reality are compatible with his mission and goals. He holds fast to a few basic standards rather than adopting those "standards" that he can get away with.
- The leader sees leadership as responsibility rather than as rank or privilege. Effective leaders are rarely "permissive." But when things go wrong—and they always do—they do not blame others.
- Because effective leaders know that they, and no one else, are ultimately responsible they are not afraid of strength in associates and subordinates. They prefer the risk of ambition on the part of able people to the risk of being served by mediocrity.
- The final requirement of effective leadership is to earn trust. To trust a leader, it is not necessary to like him or her. Nor is it necessary to agree with her or him. Trust is the conviction that the leader means what he/she says.

Let us look at two rather different kinds of leaders to put the matter in perspective: Joyce, the director of a major industrial research laboratory, and Fred, the manager responsible for a major manufacturing operation.

Joyce is hard on the managers who report to her. But she shows great consideration for the welfare of the rest of the staff. She always gives the benefit of the doubt to the person at the lowest level of the hierarchy. She is capable of understanding and is interested in every aspect of the work that goes on in the lab. When she addresses various groups of staff members, she speaks of the needs of the organization and the contribution research has made in meeting these needs. She singles out and praises outstanding contributions and contributors.

Having placed great emphasis on having the research staff understand the issues facing the corporation, she has increased the number of relevant contributions originating from the research scientists and engineers. Recognition of these contributions is most evident in the additional resources—people and equipment—made available to the individuals or groups responsible.

Such contributions on the part of a significant number of researchers have made it possible for Joyce to address major technical issues facing the corporation at the strategic level. Many researchers have been involved in the development of strategic initiatives that have had far-reaching influence upon the direction of the corporation.

Joyce never asks the corporation for additional resources for "basic research." A percentage of all new resources is "allocated" to basic research. Joyce' strategy and operating plan are the result of a bottom-up effort designed to reflect the objectives of the corporate wide strategy. Salaries are administered on merit alone. Joyce' leadership has increased the effectiveness of the research laboratory in the eyes of the corporation it serves. Significant growth and a generous budget attest to that view of her effectiveness.

Joyce is not well liked by the people closest to her, the other top-level managers of the laboratory. The scientists and engineers in the laboratory look to Joyce as the leader, the person who best understands their aspirations and who has consistently advanced their interests and offered them freedom to do creative work.

I came to know Fred, the plant manager, in the course of a task force I was asked to chair.

An information system, designed to track parts and assemblies through the manufacturing and testing process, could not be made to function as intended. This very elaborate system had been reviewed by every available expert and by two previous task forces without giving any clue to its failure mode. The second task force came to the conclusion that only concerted false input on the part of many people could produce the results which were found. This did not seem a plausible hypothesis.

Fred had the reputation of running a very disciplined operation. The performance of his plant, in terms of productivity, compared favorably with other such plants of the company. Fred was very popular with all the managers reporting to him. They knew what he expected of them and what they could expect from him. After talking to a large number of people at all levels of the operation an interesting explanation emerged.

In the course of his management career, Fred had become sold on the concept of Management-by-Objective. Anyone who *exceeded* his or her objective was rewarded accordingly. In Fred's plant, the objective was to turn out widgets; and very sophisticated and expensive widgets they were. Given the challenge to turn out more widgets than called for, Fred's very experienced group of managers found "creative" ways of meeting this challenge. The information system proved to be a hindrance; hence, the actual production control was carried out on the back of envelopes. Parts identified as scrap or rejects found their way into finished widgets. The plant consistently shipped better than one hundred percent product based on the committed parts inventory.

Both Joyce and Fred were leaders with a loyal following. Joyce led, and Fred misled. Joyce operated on the premise that she could fulfill the aspirations of her staff and serve the corporation well in the process. Fred had found *a way* of managing: Counting widgets. There was no greater challenge than to doctor the records.

Leadership is the ability to identify goals that people can identify with. Lofty goals will be better motivators than pedestrian goals. If the goals prove unattainable, the leadership will be challenged.

Abraham Zaleznik's experience (Zaleznik 1977) suggests a marked difference in the personalities which he associated with managers and leaders.

> Managers relate to people according to the role they play in a sequence of events in a decision-making *process* while leaders, who are concerned with ideas, relate in a more intuitive, empathetic way … The distinction is simply between the manager's attention to *how* things get done and a leader's to *what* the events and decisions mean to the participants.

There can be no question that the world of high technology has its share of managers who lack some of the qualities associated with leadership. My experience runs counter to the notion that management and leadership require incompatible personality traits. To be successful in today's Hi-Tech environment the organization must seek out and develop leadership qualities at all levels of management. Failure to do so is a sure way of creating the kind of "process-oriented" management style that gets in the way of creative effort.

5.6 Summary

A point has been made about management selection based upon familiarity with the qualifications of potential candidates within the organization and the potential problems of hiring managers from the outside.

We have identified the most important qualifications desired at the entry level, at intermediate levels, and at the top level of management. Concern for people and the ability to make decisions are considered paramount. Leadership is considered essential at all levels of management in a creative enterprise.

6. Managing a Project

What are some of the unique issues faced by the newly appointed manager of a project? How does one deal with these issues? Attention is given to the expectations of the project members. The use of schedules, project-tracking schemes, and reports are considered with a view to avoiding some commonly encountered problems.

6.1 The Expectations of the Project Members

You have just been put in charge of a project. Your appointment as manager of the project came about in one of several ways:

1. You have proposed and successfully sold a new idea for a project to your management.
2. Someone other than you has proposed a new project and you have been asked to assume management responsibility.
3. The project you have been asked to take over was part of a larger project. It has been split up to place increased emphasis upon some aspect of the project. As a member of the larger project, you were selected to head up this portion.
4. The person who used to manage the project has left, been promoted or fired, and you are taking over an ongoing project.

These different circumstances (and possibly others) are likely to create a range of challenges to the newly appointed manager.

> The tendency of the inexperienced manager, trained as engineer or scientist, is to focus exclusively upon the technical aspects of the project. Very likely, you were selected for this job because of relevant technical competence. Thus, you begin by doing what you understand and do best.

R. Kay, *Managing Creativity in Science and Hi-Tech*,
DOI 10.1007/978-3-642-24635-7_6, © Springer-Verlag Berlin Heidelberg 2012

Only several weeks later do you realize that nothing is happening in spite of competent people and a reasonable approach. This could be due to any one of the following reasons, related to the above-cited circumstances of your appointment:

1. You have not convinced the people assigned to the project that your approach is indeed the best. Or you have failed to make it clear to them that you are always open to their suggestions, by pointing out areas of uncertainty, thus inviting their contribution.
2. You have failed to convince the people who have proposed the project, and whom you are now supervising, that they will not be deprived of the credit for having proposed the original idea.
3. You have not convinced your former colleagues that as their new manager you will adequately represent their interests.
4. You have not convinced the people on the project that the circumstances of the former manager's departure do not reflect adversely upon the project.

How does one approach this type of problem?

The first thing to realize is the fact that, as the new manager of this group of people, whatever your relationship with them in the past, this relationship has now changed.

Like it or not, they expect you to act in accordance with *their* concept of a good manager *from the start*. They may not be able to formulate their concept clearly or to agree on a set of expectations. But this does not preclude strong reactions to anything you may wish them to do.

Hence, it becomes important for you, the new manager, to understand, as quickly as possible, the motivations and expectations of the various members of the project. This is best done on a one-on-one basis with each member of the project. It should be the occasion for information gathering, rather than conflict resolution. Once a reasonably complete assessment is available, a discussion of the situation with *your* manager should be useful. Your manager is the person (other than yourself) most interested in your success and will appreciate your readiness to ask for help in avoiding problems, rather than in fixing them.

Having arrived at some level of understanding of the group's expectations and having enlisted the advice and support of the next level of management, it is most important to provide early feedback to the group. They want to know:

* To what extent their views are shared by you and other members of the group
* What you intend to do, or not do, about the concerns they have expressed

In a meeting with the entire group you should start by emphasizing all of the recommendations you have received which you *intend to implement*.

Next, you should identify the issues which *need further discussion* before you can act upon them. For example:

Some of you have asked for the ability to sign purchase requests, rather than have me sign them. Before we can ask for a change in current procedures, can we agree on an appropriate dollar limit and how we will keep track of current expenditures?

Finally, things you *do not expect to do* should be clearly delineated with a well thought out answer. For example:

Allocation of my time will not allow me to write or rewrite your contributions to the monthly progress report. They will be passed on as your unedited input and reflect not only on your work, but on the trouble you take to present it to others.

6.2 Management Initiative

A final word of caution to the new project manager.

There is a truism as obvious as it is subtle. Namely, before developing initiative in subordinates, the manager must see to it that they *have* the initiative.

At issue here is the new manager's eagerness to be *helpful*. This is particularly true in the case where the new manager is dealing with a group of people with minimal task-relevant maturity (new to the organization and limited experience). There is a tendency for such people to come to the manager with the opener, "Boss, we've got a problem". While it is entirely reasonable for the manager to help in any way he/she can, it is essential that the person helped understand that the problem is not the manager's. Oncken and Wass (1974) suggest the following response:

At no time while I am helping you with this or any other problem will your problem become my problem. The instant your problem becomes mine, you will no longer have a problem and I cannot help a person who hasn't got a problem.

When this meeting is over, the problem will leave this office exactly the way it came in—on your back. You may ask my help at any appointed time, and we will make a joint determination of what the next move will be and who will make it.

In those rare instances where the next move turns out to be mine, you and I will determine it together. I will not make any move alone.

More serious than the new manager's eagerness to be helpful is the temptation to demonstrate superior knowledge. This will encourage subordinates at any level to shift responsibility for decisions to their manager. In the extreme case, some subordinates may be using this approach to cover up inadequacies.

I have seen Hi-Tech managers fail due to their inability to overcome the need to display superior insight. Sometimes this need is rationalized as a desire to get "to the bottom of things," or "to show a concern for the details" so as not to appear superficial.

With creative people, the fostering of *management initiative* can be even more important than the delegation of responsibility. Management initiative should not be an abstract concept or limited to management personnel. The five levels of management initiative defined by Oncken and Wass (1974) can be exercised are as follows:

1. *Wait* until told (lowest initiative)
2. *Ask* what to do

3. *Recommend,* then take corresponding action
4. *Act,* but advise at once and
5. *Act* on own, then routinely report (highest initiative)

Whoever uses initiative (1) has no control over either the timing or content of boss-imposed or system-imposed tasks, and thereby forfeits any right to complain about what he or she is told to do or when. The manager who uses initiative (2) has control over the timing but not over the content. Initiatives (3), (4), and (5) leave you in control of both, with the greatest control being at level 5.

Appendix A.1 gives an example illustrating each of the five levels of management initiative. You, the reader, may find it instructive to think of such examples based upon your own experience or imagination, before looking at this example.

6.3   The Role of Estimates and Schedules

> One of the most frequent sources of conflict between manager and subordinate is the lack of common understanding of expected results.

To ask for a particular result "as soon as possible" does not define expectations. Nor is it desirable to check every day whether a particular result has been achieved.

An experienced manager would ask for an estimate of when the result is expected. Until the new manager has developed some level of confidence in the ability of members of the group to accurately estimate the time and resources required to achieve a particular result, it is essential that the estimate be challenged and the basis for the estimate be understood, explicitly.

The agreed-upon estimate then becomes the first approximation to the expected. If there is considerable uncertainty, it may be well to check a little before the expected time of completion.

> The most effective way to develop the ability to estimate requirements, be it time or other resources, is to involve all of the people concerned in the process.

This process also serves to increase the manager's ability to rely upon others in meeting this important managerial responsibility.

Depending upon the relevant experience of the members of the project, i.e., their *Task Relevant Maturity,* the same considerations apply to the definition of the task to be accomplished. (The concept of *Task Relevant Maturity* is one I have found to be very useful. If we accept the definition of maturity as the willingness and ability to take responsibility, then this concept goes considerably beyond the notion of "experience" (Grove 1985a).)

Let the individual tell you, the manager, how many iterations are required in order to achieve a result of a given accuracy or generality.

If at all meaningful in the particular context, divide the task into subtasks, in order to get the earliest possible indication of progress or of the validity of the approach.

To restate our objective in developing schedules:

In order to find out whether our expectations are borne out by the actual course of events, initial expectations must be stated explicitly. Only by checking progress can we learn what we do well and what we tend to do poorly, and where we are inclined to over- or underestimate.

The idea of schedules may not be well received by creative people, at least initially. I have had people doing research tell me that it is impossible to schedule breakthroughs. I agree! On the other hand, regardless of how creative one's efforts, breakthroughs are the exception.

Never should schedules *be an obstacle* to the pursuit of a promising idea. The manager must convey the notion that schedules are designed to develop our ability to judge the difficulty of a task so that we may conserve that most precious and finite resource available to us, our time.

Nothing could be more negatively perceived than a manager who places undue emphasis upon schedules when judging creative effort.

In conclusion, effective, creative people have no problem accepting the notion of estimation and schedules once they understand their potential value to their *own* efforts. This value derives from the growing competence in judging alternative approaches, the hallmark of innovative people. Innovation implies the effective (timely) conversion of creative effort into results which have impact upon the field. Without such task-relevant feedback, the much professed value of experience may turn out to be nothing more than a mindless repetition of mistakes.

In most Hi-Tech organizations, internal competition for resources is keen. To become an effective competitor, it is essential that a manager develop the required skills of estimating the resources required to achieve a desired result.

6.4 Communication Within the Project

While people in general, and creative people in particular, often resent the notion of being *supervised,* they just as often show signs of anxiety if they think they are being ignored. Hence, each manager must find a natural way of communicating with the members of the group (natural, i.e., a way with which the manager feels at ease). This should allow the manager to keep abreast of the current status without being perceived as being unduly inquisitive or meddling.

What works well for many managers, besides a weekly one-on-one with each person reporting to the manager, is a weekly project meeting with all members of the project where everyone speaks their mind. Such meetings are primarily for information exchange, not for conflict resolution. When conflict is evident, the people concerned should be taken aside subsequently. The expression *let's take this problem off-line*, which comes from the notion of *online* processing, has served well in terminating discussion of issues, which are not appropriate for online resolution. Since meetings are a primary vehicle for managerial action, for gathering and providing information, and for decision making, we shall return to this topic in Sect. 10.6.

Now that the project is off to a good start, do we need the manager for anything but his/her technical contribution?

6.5 Communication with Those Outside the Project

Few projects are independent of sponsors, other projects, or support groups. As in the case of individuals within a project, there may be all sorts of mutual dependencies. These dependencies may take the form of vendors providing goods or services or other projects working on a parallel subtask. Those who need to evaluate, utilize, or distribute the project's output are dependent upon current project status information.

The manager must take the initiative of making sure that all those upon whom the project depends are able to meet their commitments—and if not, decide what to do about it.

If your project fails, it is of little consolation that it was someone else's fault.

It is also the responsibility of the manager to keep his/her manager and others who are dependent upon the project informed of the current status.

There is nothing more disturbing to *your* manager than to hear about a problem with your project from someone other than you.

6.6 Formal Project-Tracking Schemes

Some organizations have developed elaborate computerized schemes for tracking a project. If these techniques have proven themselves in the organization and hence are taken seriously by other levels of management, use them. They represent an effective medium of communication, a common language.

If you feel you need to institute such a tracking scheme of your own, be sure the need for your approach is recognized and accepted by others. There is nothing more useless than a beautiful system of project tracking which is generally ignored.

Relatively independent projects of an experimental or research nature rarely justify a tracking scheme. More development-oriented projects with much inter-dependence can only make effective use of a widely accepted, uniform system.

Whatever the origin of the system in use, be sure you understand the basis for the projections and be thoroughly familiar with the assumptions which go into them before you publicly refer to it. *There is nothing more upsetting to you as a manager than to have the presentation of a project schedule challenged, and not to have all the facts at your disposal.*

To illustrate: John's presentation of the project status included some assumptions about the progress of another group, crucial to the success of the project. Under the pressure of preparing the presentation, John had not verified the latest status of the other group.

When the assumption was questioned and John had no additional data, a telephone call revealed a major discrepancy between John's assumption and the actual status. This did irreparable damage to John's credibility as a project manager.

The people sitting in judgment had no basis for judging John's technical competence. All they had to go on was the fact that, in an important matter, John had led them to believe something, which a single telephone call proved to be wrong.

It has been argued that technical competence is a higher form of accomplishment than making status presentations. If the project fails, the people with ultimate responsibility for the consequences of success or failure can take little comfort from this. In their scheme of things, a missed delivery date for a crucial component is as significant a piece of data as is the decisive parameter in a sophisticated simulation which establishes the feasible time window for an experiment.

> What is more, once doubt has been cast upon the credibility of a manager, reestablishment of that credibility is very difficult.

A project-tracking scheme can be a valuable asset. But remember, it is the accuracy and timeliness of the data, not the elegant scheme, which makes it valuable (Oncken and Wass 1974).

The importance and ultimate value of formal project tracking are greatest for large complex projects. When such projects are primarily product development efforts, the control and tracking constitute an absolutely essential aspect of project management. While such projects can have a significant "creative" component, the subject of management and control of large projects is beyond our present scope. There is considerable literature on this subject, for example Archibald (2003) and Kerzner (2009).

6.7   Report Writing

Creative people often resent the fact that they are expected to write periodic progress reports. They feel it takes valuable time from the more creative aspects of their work. Some find report writing a painful and unwelcome task forcing them to display inadequacy. Then, there are some managers and organizations which appear to have an excessive need for reports from project personnel.

Within some Hi-Tech organizations, research and staff functions, in particular, consider reports their most visible and important output. In geographically dispersed organizations, periodic reports are often the most heavily relied upon means of communication. E-mail distribution can make reports a very cost-effective way of maintaining current awareness and satisfying a wide variety of information needs. In many Hi-Tech enterprises, report generation accounts for a significant portion of project time.

The subject of report writing not only comes up in every Hi-Tech organization, but it comes up again and again in the same organization. Given that this is recognized as an important issue, why is it not more readily resolved?

This may be due to the fact that it affects a variety of functions, whose managers do not all see the same solution. Hence, they are afraid of turning the solution over to an administrative section for fear of having an unsuitable solution imposed upon them. Yet, everyone agrees that some consistency of content detail and format is desirable.

As manager or project member, be sure you understand the objective of a particular report: who is likely to read it and what do the readers expect to gain from it. The need and purpose of reports should be reviewed, periodically. The most common complaints *on the part of recipients* is the length and the lack of relevance of the contents. *The originators* of the reports should periodically question the need for the required (or assumed to be required) amount of detail.

An occasional questionnaire verifying continued interest as well as the correct address is a must. If there is a requirement for several periodic reports, project descriptions, planning documents, proposals, etc., be sure that their relevance to your success is given a priority and that you spend only the amount of effort appropriate to the particular document.

Whenever feasible, an executive summary is a good idea. Few decision makers have the time to read the details unless they are relevant to them at the time of reading. If there is no summary, a lengthy report, at best, gets to the bottom of the *must read* file and, at worst, into the circular file. *The summary can be useful in providing you with an opportunity to point out what you consider important.* Good organization of an indexed report can overcome the problem of length when a report must cover multiple subjects or functions.

Don't let *fairness* in representing all efforts equally detract from the value of your project report. The periodic report should be designed to keep others informed of significant events affecting or being affected by your project. If there is the need for recognition of every member of the project, indeed a most necessary and worthwhile objective, a suitable vehicle should be found. A once-a-year report of the accomplishments of the group and outside contributors may be appropriate.

In large organizations, detailed reports are often required to fill the needs of staff functions and other organizational entities outside your own. Your line management must decide to what extent your time is to be invested in satisfying these needs.

You may not always feel that you have much choice in the matter of report writing. On the other hand, if you are paid to accomplish something through others—the classical definition of management—you should challenge requests which make seemingly unjustified demands upon your project resources. Have those making the demands, back them up.

6.8 Patents

One of the most valuable products of creative effort is the intellectual property represented by patentable ideas. A patented idea, first of all, protects the owner's rights: the owner may use the idea and/or may license others to use it. By refusing a license, the owner can prevent others from using the idea. In many fields of high technology, a patent portfolio can play a decisive role. For many years, the Xerox Corporation maintained its lead in the office copying field by virtue of an excellent patent portfolio. In the race for dominance in the web search market, Google's main contender lost by virtue of a lacking patent portfolio (Levy 2011, p. 89). MIT and Stanford University have been able to support long-term research from license income derived from some seminal patents in computer technology and music synthesis, respectively.

If you have failed to protect a patentable idea, you may be prevented from using your own idea which has been patented subsequently by someone else. Another important consideration derives from the fact that large organizations negotiate cross-licensing agreements based upon an assessment of their respective patent portfolios. If your firm does not have patents sought by others, you may be excluded from industry wide cross-licensing. For a high-tech start-up enterprise to go public, a strong patent portfolio is a crucial asset.

Inexperienced engineers and scientists sometimes resent the need for timely documentation and formal disclosure, as one more bureaucratic burden which interferes with their creative efforts. The rewards for significant patents can be very great. The frustration associated with seeing someone else's name on a patent relating to an idea you have failed to protect can be equally great. What is more, almost all organizations, including universities, legally obligate all members to the timely disclosure of ideas which are thought to be patentable.

In order to affect a meaningful patent policy, management must meet several prerequisites:

1. Establish an explicit patent policy and a system of rewards designed to motivate effective implementation.
2. Provide technology-relevant education and information to all new employees, preferably through a patent attorney familiar with the technology pursued in the particular organizational unit. The best exposure involves running through the process and procedures based upon a recent example from within the organization.
3. Make every scientist and engineer aware of the patent policy of the organization, including the reward system. There are a variety of ways to protect ideas, some less costly and less burdensome than an application for a patent. (It can cost more than $100,000 to have a foreign patent issued.) There is also considerable variation in the relative protection and disclosure in the case of domestic and foreign patent applications.

Note: It takes experience relevant to the particular technology involved, as well as an understanding of what is patentable, to make a judgment about the potential patent value of an idea. This judgment is difficult to acquire unless one is prepared to devote a great deal of time. Once an idea has been judged potentially valuable, the services of patent professionals are required to determine patentability. This involves a search of the prior art, based upon a preliminary set of claims defining the patentable aspects of the idea.

6.9   Summary

Attention has been called to what the project members expect from the newly appointed manager. To understand and address these expectations is the most pressing initial task. Estimates and schedules are seen as a means of establishing a common understanding of what is to be accomplished and when. The manager can fail the project as readily by inadequate communication as by inadequate understanding of the technical issues. Those outside the project who are dependent upon the results and those, upon whom the project depends, must necessarily base their judgment upon what is communicated to them.

The value of project-tracking schemes is directly related to their acceptance and the accuracy and timeliness of the data they portray.

Reports constitute a significant aspect of the creative effort in the Hi-Tech environment. Make them work for you—the need for them will not go away.

Patent documentation is an essential and important aspect of protecting the results of creative effort. Industry and universities alike no longer take this matter lightly.

7. Some Aspects of Managing a Department or Small Enterprise

The role of an explicit strategy, operating, and financial plan is considered. The elements which comprise these plans are discussed, along with their usefulness and value to Hi-Tech organizations.

As we have suggested in Chap. 5 "Desired Qualifications of Managers," the department, or mid-level manager must assume responsibility for the training and development of the entry-level manager.

The new manager, in turn, must stake his/her claim to a fair share of the department manager's attention, but without making unreasonable demands. To do this effectively, it is necessary to understand the responsibilities and priorities of the next level of management, particularly with respect to what is expected of you as a new manager.

Among the responsibilities of department managers are the periodic preparation of a strategy and operating plan. In large organizations, department managers may not be responsible for the strategy or operating plan itself, but may be asked to provide input to the next level of management for their area of responsibility. In turn, the department manager may ask the project (entry-level) manager for input.

This discussion of strategy and operating plan is also relevant to our subsequent look at the business plan required for starting an enterprise (Sect. 11.3).

7.1 Developing a Strategy

An organizational strategy should articulate common understanding as to the purpose, scope, and limits of the organization, big or small. In large organizations, the strategy sets the overall goals which provide direction for the individual units of the organization. It provides the basis for an operating plan which defines specific objectives and allocates resources.

In a small enterprise, where management responsibility falls upon one, or very few people, the strategy helps to assure that the various strategic issues are attended to, by assigning responsibility to one of the principals.

R. Kay, *Managing Creativity in Science and Hi-Tech*,
DOI 10.1007/978-3-642-24635-7_7, © Springer-Verlag Berlin Heidelberg 2012

Today, science and all manner of Hi-Tech enterprise are among the most competitive organized activities of mankind, involving a multitude of the most creative, capable, and motivated people to be found. One of the responsibilities of management is to establish a strategy appropriate to an endeavor which can be successful in this rapidly changing environment (see Chap. 14 for a discussion of the global nature and significance of this issue).

For example, a Hi-Tech enterprise may have a variety of strategic reasons to engage in research:

- To maintain or attain a position of technological leadership
- To attract top-level technical personnel
- To project a favorable image to investors
- To provide in-house consulting capability and have available a pool of unique skills relevant to the enterprise

Edward David, then President of Exxon Research and Engineering Company, on the occasion of his receipt of the Bueche Medal of the National Academy of Engineering made a strategic statement about the purpose and scope of basic research:

> In an intensely competitive world, a major strategy for overcoming the risk of missed opportunity is fundamental science and mathematics. However, I am speaking about fundamental science linked strongly to the purposes of industry… conducted by industry, or by industry consortia, or through industry–university connections in research (David 1985).

Here then is an explicit statement which defines "overcoming the risk of missed opportunity" as a major element of strategy. Any proposal which aims to implement this strategy can be evaluated in terms of the relative significance of the opportunity which it addresses. A strategy for research, designed to attract outstanding personnel, may evaluate new projects by a very different set of criteria.

A common understanding of an *explicit* strategy makes possible the broadest interpretation, and the greatest degree of freedoms in evaluating new ideas and their relevance to the organization.

This aspect of an explicit strategy constitutes its chief value. An assumed or implicit strategy may fail to provide the desired directions not only by possible inadequacies, but also by lack of general awareness.

While there may have been a traditional aversion to planning in places such as the USA in the past, the recent history of industrial research suggests a general acceptance of the necessity and value of such planning (Buderi 2000).

7.2   Elements of a Strategy

What are the elements of a strategy particularly relevant to a Hi-Tech organization?

7.2.1 Environment

A definition of the environment which delimits the scope of the organization or organizational entity to which the particular strategy applies is essential. Such an entity is sometimes referred to as a "strategic unit," generally the largest unit of an enterprise which can be encompassed by a specific strategy.

Defining the environment aims to answer questions such as the following:

- Are we responding to competitive technology or are we motivated by our desire to understand the limits of our existing technology?
- Are the ideas which we are pursuing based on the expectation of technological breakthrough or on the need for such breakthrough?
- What is the basis for our expectation of competitive advantage in reaching our goal?
- Is continued support of our endeavor based upon the expectation of maintaining, or trying to achieve, technological leadership or competitive capability? The answer to this question would influence the importance attached to basic research.
- Will our financial base, our debt structure, sustain us for the period under consideration?
- Are we seeking to improve the operation of railroads or to provide better transportation?
- Are we focused on the foundation of the organizations past success or on the most challenging and potentially game-changing issues of the future?

In addition to answering this type of questions, the strategy must include an assessment of relevant efforts elsewhere, and of expected or feasible developments which may have impact on this particular endeavor. It must include assessment of political, economic, legislative, and regulatory developments which could possibly influence the current environment.

7.2.2 Mission

A statement of mission and priorities for the organizational unit which is preparing the strategy is essential. In a large organization, this delimits the responsibilities of the division, department, or function preparing the strategy. The mission statement should define and differentiate between short- and long-term objectives. It should relate these objectives to more global objectives of the organization, where appropriate.

Very often missing from a strategy is a statement of the things which are not being done, of the competing ideas or opportunities which are not being pursued, and the reasons for their rejection. The value of defining an explicit strategy is much diminished when this element is missing.

A clear statement of mission and priorities evolves from iterative discussions which should lead all those involved to some common understanding, if not consensus. If there is a lack of consensus, this should be pointed out so that it may be reviewed subsequently.

7.2.3 Potential Impact

There is need for an explicit statement of the potential impact of the expected achievements, both upon the external environment (i.e., customers and competition) and upon the internal dependencies.

What will be the extent of the impact? Who will feel the impact, inside, outside the strategic unit? Are potentially affected units outside the organizational entity anticipating (positively or negatively) the expected results? Are there provisions for technology transfer? What are they? What is the potential impact of not achieving an objective?

To quote Edward David again:

> Even with excellent people and a sense of mission, the technology that emerges from R&D will seem irrelevant to the business unless there are tight connections between the R&D organization and the rest of the business.

In the course of the last 30 years, a statement to this effect has been accepted and reiterated by every successful director of R&D, the world over (Buderi 2000).[1]

It is general practice in Hi-Tech enterprises to develop a 5-year strategy, annually, i.e., to review, amend, and update the existing strategy. Changes in strategy are usually most pronounced in their effect upon the operating plan for the next year. In that plan, such change is reflected in reallocation of resources, based upon a new strategic direction.

7.2.4 Usefulness of Strategy

In a Hi-Tech organization, which must respond to rapidly changing technology and market developments and which depends upon creative people, the strategy takes on additional importance. Once a strategy has been established, one is in a position to develop a plan to attain specific objectives. The plan is based upon the best available ideas at a given point in time. If, however, in the course of the planning period, creative people come up with an idea which suggests a more promising objective, that idea can be evaluated in terms of its potential contribution to the strategic objectives of the enterprise.

---

[1] Much is to be learned from exemplary instances of success and failure on the part of corporate management regarding the significance of technology in a corporate strategy—Apple and Xerox (Hiltzig 1999).

From what has been said so far, it may appear that the need for an explicit strategy is confined to large organizations. This is not so:

The small enterprise with one manager or but few individuals sharing management responsibility is much more vulnerable to the lack of a well-formulated strategy. Since the strategy is the basis for an operating plan which defines areas of responsibility, it is easy for a small organization to overlook or neglect an important aspect of the strategy through failure to assign responsibility to the few people who are already doing a multitude of jobs. What is more, the small enterprise lacks the resources to compensate for error or inadequacies in people. Its financial constraints often do not allow for the time required to correct the course. This point is made forcefully by Drucker in his seminal work on management (Drucker 1985a):

> The small business needs organized and systematic management even more than the big business. In the first place, it needs a *strategy* ... The small business cannot afford to become marginal. The small business has limited resources, above all, of good people. Unless the key activities are clearly identified and assigned as responsibilities, there will be a diffusion of resources.

This is not to say that the development of a strategy, like any other aspect of management, cannot be self-defeating by virtue of overemphasis or misuse. In large organizations, it is not rare to find a readiness to reject new ideas as being "counter-strategic." Xerox's missed opportunity in the personal computer market (Dual 1983) and IBM's long rejection of the Virtual Memory (VM) operating system are examples of how vested interests *within* a large organization make use of "THE strategy" to thwart internal competition.

The process of developing a strategy has been found to facilitate common understanding about the most important issues facing an organization. In fact, it is generally held that *the process of developing* of a strategy is itself more important than the emerging document. This process is one of the most useful tools for organizational development. It provides the management team with a structured way of reviewing the organization in terms of the strategic objectives that it has set for itself. In the large organization, it helps focus upon the issue of centralized versus decentralized organization.

The process of developing a strategy also serves to avoid cross-purpose or unwanted duplication of efforts in various parts of an organization. In the small organization, it points up the need for team building in anticipation of expansion. It helps the new managers to think of their efforts in terms of the organizational objectives.

It provides top management with an opportunity to define objectives in the broadest possible terms, thus leaving a maximum amount of freedom to the component units.

The existence of a well-thought-out strategy will not guarantee success. But experience has shown that the absence of such a thought-out strategy and the absence of the planning process which gives rise to it can frustrate the endeavors of highly competent and motivated people.

In summary, we have pointed out the value of an explicit strategy, as well as the value of the process leading to its creation. Some potential liabilities have also been mentioned. The greatest liability comes from the assumption that the annual effort to produce the document will assure strategic direction for the organization. Much of the thinking reflected in that document is based upon the experience of the past. It is probably only a rough approximation of the present and contains much speculation about the future. *An explicit strategy* is *necessary, but not sufficient.* To the extent that it contains a model of the future, it need not be correct, only useful— useful in helping us recognize the value of new information and insights. In today's rapidly changing world of science and technology, one of the more demanding tasks of management, at all levels, is to be ever alert to the "weak signals" which are the harbingers of change in our strategic outlook.[2]

7.3   Developing an Operating Plan

The purpose of the operating plan is to define the manner in which each strategic unit is to meet its strategic objectives during the next planning period. This is usually 1 year, and rarely more than 2.

The operating plan does this by stating what is to be done, how, and by whom. It is important that it includes the milestones which will serve to measure the progress toward the objectives.

Most medium or large organizations have developed a format for such a plan, designed to achieve consistency of detail and facilitate ease of understanding. At the lowest or project level, a typical format may look something like this:

- Name of project; name of manager; planning period
- One paragraph describing the major activities and objectives of the project
- Two or three paragraphs giving relevant recent history and accomplishments
- Milestones (5 or less); brief statements of specific objectives to be accomplished with dates
- Resources on-board or to be acquired during planning period
- Manpower: names, skills, area of responsibility, on-board, or expected date of arrival
- Special equipment or facilities

At the next level, which may be the top management of a small business or the functional level of a large organization, such as a product group, the sales organization, or a manufacturing location, the plans of the individual subunits or projects are aggregated in the most relevant fashion. Here, the objective of the plan is to provide an overview of how resources are devoted to the key objectives of the strategic unit, interdependencies, and key dates which will allow top management to be aware of progress toward the stated objectives.

---

[2]For a uniquely instructive example of the role of strategic thinking in the turnaround of IBM in the 1990s, see Gerstner (2002).

The financial plan or budget, which usually dictates or at least puts constraints upon the operating plan, is often separated from the operating plan, particularly when the operating plan involves separate or loosely connected subunits. That is, the operating plan is a bottom–up effort which should be subjected to the top–down allocation of resources in an iterative process.

In a small organization where management responsibility falls upon a few people, responsibility for the operating plan and financial plan must be shared by the same people. Here the strategy must define the basis for reconciling conflicting demands, such as the need for curtailing operations vs. the need for additional financing through borrowing, etc.

It is important for new managers to become familiar with the process and the specific procedures that are followed in the organization. Managers can just as easily fail their project, by being on a trip during the plan finalization period, by being tardy with their input, or, indeed, by having a poor plan.

A sample project operating plan is included in the following pages.

Sample Project Plan

PROJECT PLAN: R&D LABORATORY, BUILDING 404
Natural Sciences Division, John Worley, Manager

Project No.	Project Name	Manager	Phone
4711	Laboratory Automation	Pat Manning	7451

Planning period: 1 Jan 2012–1 Jan 2013

Activities and Objectives: The Lab Automation Project develops advanced computing functions for the Natural Sciences Division of the central R&D laboratory. Emphasis is on real-time systems and integration with all related computing facilities. The objective is an environment which provides the tools and facilities for individual users to develop their applications with a maximum of independence and provides standard interfaces to the existing network functions.

The project started in 2008 in the central computing facility to meet the unique needs for real-time system users.

In 2010, the project was transferred to the R&D laboratory, to better meet the advanced requirements of very high data rate, long-running experiments.

By mid 2009, a comprehensive approach to data acquisition, storage, and subsequent processing allowing both stand-alone operation and networked access to the central facility was approved by all the concerned functions.

This capability is largely implemented and has served as a model for other parts of the organization having real-time computing requirements.

Objectives for the Planning Period

1. Transfer of the system to the DataComp subsidiary for further development and marketing. Transfer agreement by 5-1-12 (see item 2).
2. Agreement on priority of enhancement to be provided by DataComp (by 4-1-13).

3. Preliminary design of an advanced graphics capability for concurrence of all interested groups in the R&D laboratory by mid-year.
4. Implementation and test of a newly conceived algorithm for upside-down Schlemmer transforms, widely used throughout the laboratory. Mid-year 2012.

People on Board: YE 2011 (5P), (2T), (1A)*) YE 2012 (6P), (4T), (1A)
Hieronymus Bush, System Programming, Graphics Implementation (T)
Olive Goldberg, User Consultant, Real-time Systems (P)
Betty Green, Secretary/Program Librarian (A)
Tony Hu, Temp. Assign. from Math. Dept., Algorithms, 6-12(P)
Jill Jepson, System Programming, Graphics Architecture (P)
Wanda Lanovski, System Programming, Real-time Systems (P)
Pat Manning, Manager, Graphics Architecture (P)
Fred Overlook, Programming, Real-time Systems (T)
Martin Spitz, System Programming, Network Protocols (P)
( ) System Programming, Graphics Implementation (T)
( ) System Programming, Graphics Implementation (T)

Special Equipment: IBM XYZ (2); FP-180 Hypercube; Requested for Delivery 6-12: AGS-2000 (4)

---

*) (P) Professional, (T) Technical, (A) Administrative.

7.4 The Financial Plan

Understanding and appreciation of the subtleties of a financial plan are generally outside the experience of technical and scientific personnel. This is not because mastering this subject is unusually difficult, but because of a general reluctance on their part to invest the time required to acquire the necessary skills.

It is a fallacy to assume that acquisition and development of these skills are trivial, in terms of the time and intellectual effort required. A Ph.D. in economics is not a prerequisite for the chief financial officer of an organization. Yet, management failure in the realm of financial planning is probably the most frequent cause of setbacks. It is noteworthy that the financial success of Google was crucially dependent upon the second-price auction model of its advertising ranking strategy (Levy 2011, p 99).

It turns out that it is easier to acquire financial management skills than to learn to apply them to the wide variety of scenarios encountered.

One may compare this to acquiring the skill of reading music and applying this skill to the conducting of an orchestra.

Lou Gerstner's account of IBM's financial fall and rise in the 1990s provides insight into the complexity and scope of such problems at the CEO level (Gerstner 2002).

We return to the subject of financial planning under the heading of "Financial Controls" (Sect. 11.4) and "Evaluation of Proposals" (Sect. 12.6).

7.5 Summary

The purpose and value of a strategy and operating plan and the elements of such plans have been presented. A point was made of the importance to both large and small undertakings. The relevance to creative efforts and the value of the process of developing a strategy and operating plan were stressed. Attention was called to the tendency of scientists and engineers to underestimate the significance and complexity of financial planning.

In Sect. 9.3, the discussion of the evaluation of R&D considers the implications of technology-, market-, and finance-driven strategies.

8. Managing Creative People in the Hi-Tech Environment

In the preceding chapters, the management of people has appeared as a common thread. At this point, we will focus upon some specific aspects of people management: recruiting of creative people, performance evaluation, compensation, and management style.

8.1 Recruiting

A unique concern of the world of Science and Hi-Tech is the management of creative people. Here we will try to translate this concern into concrete terms.

We start with recruiting, one of the most important investment decisions of management, and something which requires decisive action on the part of the individual seeking a position.

To put the matter in perspective, assume a starting salary of $80,000 and an annual rate of increase of 5%; in this case, salary alone represents a more than $2.5 million investment over a period of 20 years. If we take the figure of $200,000, commonly used as annual budget for a Hi-Tech professional, this becomes $6.5 million!—How many $6.5 million dollar investment decisions are Hi-Tech managers called upon to make? What is more, it is difficult to quantify the missed-opportunity-cost of hiring anyone but the person best matched to the job.

8.1.1 Recruiting: The Applicant's Position

Recruiting the most creative, advanced-degree Hi-Tech graduates places the burden upon the recruiter since the competition for such people is always very keen. On the other hand, prospective graduates are interested in optimizing their choice among the available opportunities. This requires an effort on their part, starting at least a

R. Kay, *Managing Creativity in Science and Hi-Tech*,
DOI 10.1007/978-3-642-24635-7_8, © Springer-Verlag Berlin Heidelberg 2012

year before the date of the expected advanced degree: an effort to know about and be known to the organizations in which they are interested.

No enterprise can be counted upon to have a requirement for your talent at precisely the point when you are ready. But once your ability has been recognized, your future availability can be anticipated and a position or "slot" may be set aside. If this is the applicant's first experience of the interviewing process, there is a good deal to be learned. If the applicant has very definite ideas about the desired type of work, this is an excellent opportunity to find out how potential employers *currently* value this particular area of activity. If, on the other hand, the applicant does not have well-formulated interests, this is an opportunity to find out about the range of possibilities open to someone with a particular background and set of skills. It is instructive for the applicant to ask various members of the organization about the role of the creative person in the particular setting for which they speak.

"Who are the people perceived to make decisive contributions, the idea people or the implementers? Is the emphasis on team or individual effort? What have been the contributions of the particular group that have earned recognition? What is the role the applicant is expected to play, now and in two years from now?"

These questions go beyond the purely technical nature of a job.

8.1.2 *What Management Potential?*

One of the less obvious qualifications the recruiter may be looking for is the applicant's management potential; if not the recruiter, then some of the people the applicant will meet during an on-site visit. While you may have done little, if any, explicit managing during your academic career, experienced managers or recruiters can learn a good deal about your management potential. And so can you!

- Your interest and record in assuming leadership in sports, student activities, etc.
- Your choice of nonacademic activities tells something about you. To become an effective manager, it helps to have learned to appreciate the need for taking direction. Mountain climbing and sailing, as well as most team sports, provide this experience. Tennis and skiing are less likely to do so.

New graduates, who are about to make an important commitment in accepting their first position, should be concerned about the criteria the organization uses in management selection. Subsequent career decisions, leading to a change in position, are likely to be more constrained by personal factors beyond the job itself. The person may have found roots in a community, acquired family obligations, or developed a circle of friends at the place of work, factors which deter change. Hence, a positive first experience with a well-established and well-managed organization may be desirable. A relatively unproven start-up organization, on the other hand, may provide greater opportunity for early and varied management experience. In the former case, one learns slowly, by the example of others. In the start-up organization, one learns relatively fast by personal experience, even if sometimes painful, since the individual's mistake is more likely to have a significant impact.

8.1.3 The On-Site Interview

Certainly the way you, the applicant, conduct yourself and react to your on-site interview tells something about your sensitivity to management issues.

- Have you prepared by learning something about the organization you are about to interview? Do you have some relevant questions?
- Have you thought through a set of criteria by which you are going to evaluate possible offers?
- Have you ascertained the makeup of the audience who will listen to the presentation of your thesis work? (More about this in Sect. 10.1 where we deal with the subject per se.)
- Have you ascertained before starting your round of interviews the position and background of the people you are going to meet?
- Do you understand the day's schedule and what to do if something goes wrong? There is little consolation to know that it was someone else's fault, when you could have done something to anticipate the situation.

The interviewing organization is making a significant investment decision about hiring you. You are making a decision which is likely to have a significant effect upon your career. A flippant or casual attitude suggests lack of maturity or poor judgment on your part. It may be perceived as a sign of insecurity.

Shortcomings on the part of the interviewing organization are more difficult to assess. Based on a single experience, it is usually not possible for the applicant to conclude that this experience is representative of the entire organization or only reflects upon the individuals involved. This is one of the reasons why it is important that the person interviewed should have the opportunity to talk to potential colleagues and at least two levels of management.

8.1.4 The Recruiter's Position

Google's staffing manager, a guy who deals with 3,000 résumés daily looks for raw intellect, learning agility, diversity, leadership and innovation (Wang 2011).

Given the fact that, for both sides, the recruiting process has long-term implications, it is surprising that often it is not taken more seriously. There is at least one plausible explanation for this. Even with the best intentions and preparation, and with the involvement of experienced managers and personnel professionals, there is no way to guarantee a wise decision. Every experienced manager can recount some spectacular instance of how his or her judgment has failed to stand the test of time.

The most useful guideline for the manager recruiting professional people is to bank on the intelligence and self-interest of the applicant. Provide the applicant with honest and relevant information about the assignment, the organization, and possible career path. Make it clear that the ultimate decision is that of the applicant. The interview process should be designed to help the applicant make that decision.

A frequent ingredient of poor judgment is the short-term view of the manager who has a position to fill. In looking for creative people, the prime concern should be the long-term potential of the applicant. All too often, the mastery of skills or unique knowledge relevant to the problem, foremost in the manager's mind, takes precedence over independent thinking, eagerness to learn, drive, initiative, and management potential.

This is another reason to involve higher levels of management in the interviewing process. I believe that the importance an organization attaches to the recruiting process correlates well with the success rate in hiring (and retaining!) good people. Turnover relative to similar organizations is one of the accepted measures. Even in an organizational unit of 500–1,000 people, the director should be involved in the selection of all professionals. A 5% annual growth would mean spending 1 h with each of 25 people per year, assuming that professionals make up one half of the new staff. The person with ultimate responsibility for the success of the organization is the one who can most effectively convey the importance the organization attaches to its people. That person's involvement should not appear as one of final approval, but one of making sure the process of selection is working as intended.

Often, recruiting of professionals is done by people trained in the same academic discipline as the people to be interviewed. As a consequence, some organizations fail to place adequate emphasis upon management potential as criteria for hiring scientists and engineers. (This may account for the negative reaction to management by newly hired scientists and engineers referred to earlier.) Inadequate emphasis upon management potential is likely to become a limiting factor in the evolution of the organizational unit. The smaller the unit, the more difficult it will be to overcome this limitation. In a Hi-Tech undertaking with a significant creative component, the need for change makes equally great demands upon management as the desire to grow and hence the emphasis upon evolution, i.e., not only growth.

It should *not* be concluded from the above that management potential is to be regarded as the overriding factor in the selection of scientists and engineers for an environment where creativity plays a part. In fact, to settle for anything less than the best trained and qualified people is to accept a decisive handicap in today's Hi-Tech world. To emphasize the implication of the "best trained and qualified" I would like to refer to a Harvard Business Review article (Quinn et al. 1996) in which the authors define four levels of an organization's professional intellect, presented here in order of increasing importance:

Cognitive knowledge (or know-what) is the basic mastery of a discipline that professionals achieve through extensive training and certification.

Advanced skills (know-how) translate "book learning" into effective execution. The ability to apply the rules of a discipline to complex real-world problems is the most widespread value-creating professional skill level.

Systems understanding (know-why) is deep knowledge of the web of cause-and-effect relationships underlying a discipline.

Self-motivated creativity (care-why) consists of will, motivation, and adaptability for success.

Before leaving this topic, a word about the recruiting of support personnel. The success of any scientific or engineering enterprise depends to a large measure upon its technicians, programmers, secretaries, and other supporting skills. With most managers coming from the ranks of professionals, the supporting staff does not always get adequate attention.

Failure to meet the expectations of a competent support staff is likely to create a poor attitude. This will stifle the initiative of the very people who are the first to encounter a problem: The technician observing unexpected results from an experiment in progress or the secretary getting wind of a developing personnel issue within the department.

In a predominantly professional environment, it is essential that salaries and working conditions of the support staff be such as to attract and keep the right people. Every effort must be made to avoid the "second class citizen syndrome." Symptomatic of the syndrome is the technician who does not have a desk of his/her own or the secretary whose desk occupies the position of a traffic island.

What are the characteristics of the "right people" in the Hi-Tech environment? My unqualified answer: Basic intelligence, defined in terms of ability to learn. It is more important for a secretary in such an environment to have shown aptitude and interest in a high school science class, than exceptional speed or accuracy in typing. In the course of a career, a technician needs to learn a wide variety of new skills. I have seen machinists learn to operate an electron microscope at age 45, as well as laboratory technicians become programmers late in their working life.

In some organizations, entry-level secretaries are hired by supervisors in some administrative function. This is likely to eliminate the people with the greatest potential for effective contribution to the Hi-Tech environment. Since many scientists and engineers lack administrative skills, every secretary should be looked upon as a potential administrative assistant to a manager. Every technician or programmer should be evaluated as potential manager of a support function.

Given the fact that it is not possible to predict the future development of an individual, not all of the people to be hired will meet the highest expectations. But to look only for the specific skills, which are immediately relevant, is likely to sow the seeds for future personnel problems.

Given the high rate of change in the world of science and high technology, I would go so far as to generalize the primacy of learning ability for every job in such an environment. Based upon this premise, some Hi-Tech organizations have been able to retrain many thousands of skilled people over the years. This not only provides the retrained people with new challenges, but also produces a uniquely skilled and motivated workforce with organization-specific knowledge and skills.

If the right people for the job have been recruited, it is incumbent upon the professional or manager immediately responsible to assure that support people are sufficiently challenged and adequately recognized for their contributions. To thank a secretary or technician in front of the assembled group for having pointed out a problem to you will be as rewarding to the secretary or technician as it will be to the person giving the praise.

8.2 Performance Evaluation

No aspect of the management of creative people can play a more significant role than the evaluation of individual performance. We shall try to understand the basis for this assertion. To the discussion of effective evaluation procedures, we add some observations about wanted and unwanted consequences of performance evaluation. Opinion surveys are suggested as an organizational performance evaluation.

8.2.1 Why Performance Evaluation?

One of the things a prospective employee wants to learn about the organization she or he is interviewing is the manner in which personal performance will be evaluated and by whom. Very likely, pay increases and promotions will be based upon such evaluation.

Lack of a well-established procedure for formal performance evaluation (or as it is sometimes called, periodic appraisal) can be troublesome. Without a formalized procedure, it is difficult to assure higher levels of management, and for that matter, assure the individual that an evaluation has taken place. By "formal" we mean, something that is put in writing and occurs at fairly regular intervals. If there is no written appraisal, it is difficult to tell what has been said and what has been intended; what you, the person appraised, have been told and what you have heard. If one thinks about verbal communication in what may be an emotionally charged atmosphere, one realizes how easy it is for either party to develop an attitude early on in the discussion, which dominates and reinforces negative feelings instead of providing the basis for constructive interchange.

When all is well and you have been rated a star performer, it is unlikely that there will be a problem. But early in your career, you rarely have a realistic understanding of what represents star performance in a particular organizational unit. Prior to your first job, you may have been a straight A student and always managed to be among the top of your class. What a surprise to find that you are now being evaluated relative to others, who not only have similar credentials, but also have considerably more experience.

If the performance evaluation is on the whole positive, both the manager and the person appraised may feel that it was superfluous. This is fallacious reasoning: For some people, including outstanding performers, waiting for a tardy appraisal can be the cause of great anxiety. What is more, the need for praise and recognition is one that is never fully satisfied. This is the reason why formal recognition, a form of praise before your peers, is such an effective kind of reward.

Evaluation of people is such an intensely personal matter that efforts to teach the "appraisal process" via management training are not very effective. Since this is one of the more important nontechnical responsibilities of a manager, one's own appraisal provides a most valuable experience in preparation for management.

8.2.2 The Evaluation Process

Performance evaluations are most effective if they are based upon a "performance plan" that is put in writing at the beginning of the period to be covered by the appraisal. This plan is worked out between the person and the evaluating manager. It essentially outlines the activities which that person will be engaged in for the projected planning period and the results that are expected. In the case of the experienced professional, it may be desirable to have the individual to be evaluated, rather than the manager prepare the first draft of the plan.

The appraisal then evaluates performance relative to a prior plan which you, the person appraised, and your manager have agreed upon. It is important to note that Performance Evaluation is not a rating of ability. There are many reasons why your performance may or may not correlate with your ability, for example:

- You may be a person of outstanding ability, but have been prevented from applying yourself to the assigned task for a variety of reasons.
- Your capability may be a matter of past record, but you did not apply yourself to the task agreed upon.
- Your performance may have been rated below average in the past, but a more suitable assignment has allowed you to perform in accordance with your true ability.

A good manager has not only talked to you about specific problems before the appraisal, but presumably has tried to help you overcome these problems. The time of the formal evaluation meetings—let alone the written appraisal—is not the time to first learn about a problem. Rather it should be an occasion to refer back to previous discussion of the issue, maybe a question about the progress in solving the problem, or commendation about the progress made.

It is therefore good practice for the manager to present the written appraisal sometime before the scheduled face-to-face meeting. It gives the person to be evaluated an equal chance to prepare for the occasion. It reinforces the feeling of shared responsibility for making this an effective means to performance improvement.

A typical appraisal document (see example on following pages) will list several independent tasks which reflect the activities of the evaluation period. They may be listed in order of importance with an indication of the weight attached to each task at the time of planning. The planning part of the document defines the task to be accomplished; the evaluation part comments upon the degree to which the task was completed and the quality of the result. Each task is rated as "Exceeds requirement," "Meets requirements," or "Does not meet requirements." Other designations providing for a scale larger than three may be used. My own experience suggests that designations such as "Far exceeds requirements" or "Unsatisfactory," in addition to the above three, create more problems than they solve.

The summary rating is frequently used as a basis for periodic salary adjustments, a subject discussed in Sect. 8.3.

Some additional features of the document suggest some other important aspects of the evaluation process. There is an explicit space for unplanned activities.

The example also suggests the importance attached to the initiative on the part of the creative individual.

Some organizations make nontechnical contributions and attitudes such as help extended to others and contribution to a pleasant working environment, an explicit part of the evaluation.

The signatures and any meaningful comments reflect the value attached to the appraisal process by all involved. When appraisals are reviewed subsequently in considering promotions or reassignments, their usefulness is limited if the comments on the appraisal are perfunctory. This is much like a personal letter of reference which lacks specific content.

The appraisal document becomes part of the individual's confidential personnel record which is destroyed after a period of time, usually 3–5 years. Access to these records should be controlled by the manager of the individual and should be based on the "need-to-know" only.

8.2.3 Observations About the Evaluation Process

It is a good idea for managers to review appraisals with the next level of management. This is a way of ensuring that this important management task is taken seriously. It provides the basis for evaluating a manager in the performance of this part of the job. It also means that responsibility for personnel decisions is shared by at least two levels of management.

Since the appraisal also reflects upon the performance of the manager, she/he is more disposed to stress the positive rather than the negative aspects of performance. This is appropriate since formal appraisal should be a constructive interaction, designed to benefit the appraised person and improve the person's future performance.

While it is desirable to start and end an appraisal on a positive note, if a serious problem exists, it necessarily will dominate the appraisal. Failure to address such problems is a disservice to the person appraised and to the organization. The relative emphasis of an appraisal session should be carefully considered in advance. Most people are limited in the number of messages which they can deal with at a given time. A negative comment about a person's performance can cause that person to tune out anything positive which may follow. Serious problems which are reflected in a person's performance must be recognized and addressed in a direct manner. What is the basis for such problems?

- The creative person may be frustrated by the need for time-consuming implementation of an idea, a sometimes tedious task which preempts other creative work.
- A setback due to the failure of an approach which looked promising at the outset can be a demotivator.
- Learning of recent competing developments, which may upstage work which has been in progress for some time, can be a serious blow.
- Personal problems of the individual, unknown to the manager, may adversely affect performance.

Hubert had always been recognized and respected for his creative solutions to a variety of technical problems. He was also a very productive and cooperative member of the group. Outgoing, cheerful, considerate, and friendly would be the way to describe him.

From one day to the next, Hubert was a changed person. He sat in his office distraught, staring into space, largely inactive.

Given the very positive and open relationship between them, Jim, the manager, expected Hubert to take the initiative and come to him with his problem. When this did not happen, Jim became concerned that he may have caused Hubert's problem. Jim faced the dilemma of either seeming indifferent to the distress of a valued member of the group or inappropriately intruding upon Hubert's private affairs.

To resolve this issue, Jim went to ask Hubert if he had in any way contributed to his apparent distress and if he could do something to correct the situation. Hubert was relieved to have Jim take the initiative. He indicated that he had not been able to bring himself to talk about a very personal matter: his wife had left him, without him as much as suspecting a problem.

While the manager may not be able to alleviate the problem, she or he must show understanding and encouragement. Outstandingly creative people, by definition, perform outside the norm. It is not unusual for such individuals to react in ways which would be considered extreme when judging "average" people.

Performance Planning

Responsibilities (Key words to describe the Major elements of this employee's job.)	Results to be Achieved (A more specific statement of the employee's key responsibilities and/or goals employee can reasonably be expected to achieve in coming period.)	Relative Importance
Surface Spectroscopy	Responsibility for theoretical and experimental work relating to surface spectroscopy. Advancement of state-of-the-art.	10
Implementation	Completion and installation of spectrometer modification by model shop.	10
Validation	Calibration and validation of predicted sensitivity.	10
Publication	Timely publication of results.	4
Technology Transfer	Determine feasibility of in-house use.	7
	Determine possible value outside RBI	4
Optional Growth Assignment (Special work assignments, educational or professional activities undertaken to improve and augment the employee's job related capabilities.)		
The far reaching implications for safety procedures motivated the high priority for this task, now that the method has been validated.		10
Changes In Performance Plan (May be recorded anytime during the appraisal period.)		
Feasibility study of using transforms to speed up data reduction for Quality Control Dept.		5

Performance Evaluation

Actual Achievement.	Level of Achievement	Exceeded	Consistently Met	Did Not Meet
During this planning period Janet completed the implementation of her ideas leading to a significant improvement of the sensitivity of measuring surface contamination.				
Janet managed to get excellent cooperation from the model shop in the fabrication and installation of the spectrometer attachment.			x	
The calibration and validation demonstrated conclusively the improved accuracy (nearly two orders of magnitude) which was predicted by what has been recognized as a very elegant and convincing extension of the theory.		x		
Judicious preparation and preliminary clearance made it possible to submit the paper one week after completion of the validation work.			x	
Janet enlisted the help of a number of people in identifying product groups which could provide a convincing demonstration of the value of the method to RBI. She is now receiving support from 3 product groups interested in obtaining attachments for their own instruments. They have taken responsibility for preparing a complete set of specifications and drawings.		x		
Janet took the initiative of enlisting the legal department in negotiating a licence agreement for the instrument manufacturer to facilitate early availability of this capability to the scientific community at large.		x		

Additional Significant Accomplishments

Janet identified the significance of applying this method to monitor otherwise undetected toxic materials, widely used in industry. She has given a number of talks about this to those most likely to benefit from such a monitoring system.		x		
With the approval of the dependent department, Janet initiated transfer of responsibility for this commitment to the central scientific group, with Janet consulting as needed. This could have been done in a more timely fashion.				x

Continuing Responsibilities

(Responsibilities, not covered at left, to be considered only when they have had a *significant* positive or negative effect on the overall performance.)

Relationships with Others (Job Related)

(Significant positive or negative influence this employee has had on the *performance* of other RBI employees.)

Janet's enthusiasm, high standards and commitment have motivated others involved in project

Overall Rating

(Considering all factors, check the definition which best describes this employee's overall performance during the past period.)

Satisfactory

☒ Results achieved **consistently met** the requirements of job **and exceeded the requirements in many areas.**

❑ Results achieved **consistently met** the requirements of the job.

Unsatifactory

❑ Results achieved **did not meet** the requirements of the job.

Counseling Summary

Employee Strengths	Suggested Improvements
1. Technical competence	1. Tends to overextend; better planning
2. Goal orientation	2.
3. Breadth of interest	3.

Significant Interview Comments

(Record here only those additional significant items brought up during the discussion by either you or the employee which are not recorded elsewhere in this document.)

Janet has only recently started to give thought to her long term objectives. She would welcome the opportunity to be exposed to some management training, but is not receptive to an extended staff assignment toward that end. She has been able to increase her analytical skills through independent study.

Frances Calabria	Frances Calabria	
Manager's Signature	Print Name	Date of Interview

Employee Review

Optional Comments: If the employee wishes to do so, any comments concerning the performance plan or evaluation (for example, agreement or disagreement) may be indicated in space provided below.

There are several ideas which I would like to pursue, and realize that this can only be done through others. I am therefore prepared to assume management responsibility, but would like to do better than learn through trial and error.

I have reviewed this document and discussed the contents with my manager. My signature means that I have been advised of my performance status but does not necessarily imply that I agree with this evaluation.

Janet Wilson	
Emloyee's Signature	Date

Management Review

Optional Comments

Janet should be considered for the next Career Development Workshop. I am very much impressed with her recent achievements.

RD	Roger DeWalt	
Reviewer's Signature	Print Name	Date

Not only setbacks but also sudden success or recognition can be very unsettling. I have observed seemingly radical changes in personality upon unexpected promotion or recognition.

This concern may seem unwarranted. It would appear that exceptional performers could well take care of themselves. Experience tells a different story. Outstanding, creative achievements are often the result of intense application of every personal resource at one's disposal, not the least of which is time. Frequently this means the exclusion of other interests and obligations. An extended vacation with the family may do more for the long-term stability of an outstandingly creative person, than the seductive glory of invited papers in faraway places or promotion to a job with increasing demands upon the person's time.

Exceptional performance is often achieved by the pursuit of a very narrow goal over an appreciable period of time. The exceptional performer has thus been forced to neglect broader aspects of his or her field, which impairs the ability to embark upon another worthwhile pursuit. This may result in the attempt to relive past glory by trivial elaboration of an outstanding idea whose time has passed.

One of the imperatives of managing creativity is that management attention be concentrated upon the top performers. There is no better measure of future potential, than demonstrated results. While the bottom performers are likely to make the greatest demands upon a manager's time, this time is not likely to produce commensurate results. What is more, neglect of the top performers will surely give everyone the wrong message. And the brightest people are often most able to benefit from the experience of others. It is important to realize that creative individuals may not be disposed to seek out or accept help. This does not mean that they do n't need it.

It is not uncommon for the author of an idea to worry about losing control of its implementation. Once the idea has been recognized to justify additional resources, the author of the idea may not choose or be chosen to head up such an effort. In a similar vein, the euphoria of a widely acclaimed accomplishment is often followed by deep self-doubt about one's future. Outstanding performers know that coasting is a downhill process. On the other hand, their criteria for future achievement may be unrealistic.

The problems alluded to are among the most challenging faced by the manager of creative people. Only those who have succeeded in establishing a supportive relationship can hope to deal with such problems effectively. The ongoing task of performance evaluation is a vehicle for building such relationships.

8.2.4 Consequences of Evaluation: Wanted and Unwanted

"You exceeded the plan in all areas. Bravo!"

Alternatively, you have not achieved some or any of the expected. The appraisal session should, at the very least, increase your understanding of how your manager looks at your performance. You may not always agree, but it helps to understand the basis of disagreement and this hopefully leads to resolution. Resolution does not necessarily mean agreement. Resolution means a commitment to a course of action, defined by the manager. Since there is always room for honest disagreement, the manager's responsibility in resolving such issues must be recognized.

An altogether negative evaluation clearly requires some more drastic corrective action. If you and your manager cannot agree upon a course of action appropriate to the situation, you may want to talk to your manager's manager and get his/her point of view. There may be a personnel manager in the organization who may be able to help you understand what action to take. Generally, the evaluating manager should suggest the appropriate recourse to the staff member.

There are divergent views about the efficacy of motivating presumably capable people who are underperforming in an environment where creative accomplishment is the raison d'être. The competitive nature of such effort leaves little leeway for distraction. Large industrial organizations can usually find appropriately less challenging assignments. There are those who hold the view that no effort is too great to rehabilitate a once effective contributor. In my experience, they often succumb when the competitive pressure becomes large enough or their own effectiveness is directly impacted.

Another potential, if rare, by-product of the evaluation process is that of someone threatening to resign. This is a course of action which is appropriate only if one is prepared to leave that very moment. The manager who has been so threatened by you does not have many alternatives:

- The manager accepts your leaving with the idea of getting rid of someone with whom he/she has not been able to get along. This has to be explained to higher management and may reflect poorly upon the manager. If the manager feels justified in accepting your leaving, you have saved the organization the trouble of firing you, which can be an unwelcome and sometimes drawn-out, costly procedure.
- Alternatively, the relationship with your manager has been further strained since an accommodation must be found in the face of a threat, something very few people do gracefully.

An experienced manager would first of all offer to ignore the threat, presumably made under emotional stress and offer to meet again the next day, when cooler tempers prevail. If you were a valuable employee, the manager would immediately enlist the help of his/her manager or personnel department to persuade you to reconsider.

Resolution of conflict by threat, in general, is to be avoided. It suggests lack of willingness or ability on the part of the threatening person to deal with the situation in a constructive fashion. A threat is a form of abuse which many people cannot tolerate. It well behooves an organization to have developed a generally understood set of guidelines for accepted behavior. (More about this subject under the heading "Organizational Culture" in Chap. 13.)

There are situations which warrant resignation, i.e., other than an employee's decision to take another job. Resignation is sometimes the only way out if an employee is personally harassed or assaulted, physically or verbally, or asked to participate in a criminal act or an act which violates the ethical standards of the employee or of the organization. In such a situation, the employee may want to appeal to higher management, in order to protect vested rights or other accrued benefits.

In many parts of the world, it is a right of the employer to request a person to resign or to "effect involuntary separation," the euphemism for firing a person. The employer may be challenged to show good cause and, of course, cannot be discriminatory in firing people.

For professional people in particular, being fired is a very traumatic experience. Most organizations have a well-established procedure of initiating and carrying out involuntary separation. Good personnel policy dictates that this procedure be adhered to, to protect employees from arbitrary decisions on the part of a manager and to protect the organization in the case of subsequent attempts at legal redress. These procedures involve counseling and, when appropriate, the offer of medical assistance.

The offer of medical assistance is a particularly sensitive matter. Only a qualified medical practitioner is in a position to identify the need for counseling or treatment. Hence, it is wise for the manager to seek such qualified assistance.

When an employee's performance no longer meets minimum requirements, the procedure sometimes calls for the performance of very specific assignments to be completed in a specified period, before the decision can be made to fire someone.

It well behooves the new manager to become familiar with all these procedures, since inappropriate action can foil the best intentions.

Before leaving this topic, let us bring up the matter of personal friendship between the manager and the person to be appraised. If either person feels uneasy about this, the manager would be well advised not to have such a friend in the group. Having always attached very high value to personal friendship, I would avoid any situation which could put such a relationship in jeopardy. This is a very personal view, which I would not wish to impose upon others. There are numerous examples of outstandingly successful business relationships between close personal friends. If you are not sure about your own feelings in this matter, ask yourself if you could conceive of a likely situation which would cause you loss of sleep or some other noticeable discomfort.

In closing the subject of performance evaluation, I feel compelled to share an incident from my teaching experience.

Joe had been one of the older participants in a management workshop conducted at a graduate school of engineering. We met several months later and the conversation came around to the subject of performance evaluation. Joe surprised me by his total rejection of evaluating others or being evaluated. He felt strongly that this was incompatible with his idea of professionalism. Evidently, the intended message had gotten lost in the discussion of the fine points of the process.

The need for formal evaluation is based on the observation that every interaction between members of a group and their manager can be taken as an implied performance evaluation, whether intended or not. An innocent "How are you doing?" may lead the group member to think, "You must be questioning my performance, otherwise you would not be asking." A colleague's salary increase or promotion cannot help but make you wonder why you have been passed up; presumably, management makes such decisions based on an evaluation of relative performance. Employees have a right to know where they stand. To deny that right is to appear arbitrary and not accountable. Many an owner–manager has had to learn the consequences of "I am not accountable to anyone" the hard way.

> There is general agreement among experienced managers that formal performance evaluation is the most important and effective task-relevant feedback. It is the basis for a trusting relationship and the most readily accepted procedure for taking corrective action and it provides a basis for salary administration.

Our detailed discussion of performance evaluation is intended to bring out the subtleties of this important subject. Rather than a rigid prescription of procedure, it should be taken as a descriptive example, to be adapted to a given set of circumstances. It should be subject to modification as these circumstances change.

A discussion of the relative virtues of a three- or five-step scale of performance evaluation is largely irrelevant. What is important is that there be a good reason for choosing one over the other, one which fits the situation.

8.2.5 Opinion Surveys

An opinion survey can be a useful device for obtaining a periodic evaluation of an entire organization, particularly with respect to how it performs in the area of people management.

Anonymity of the individual response is essential. To maintain anonymity and still be able to associate a set of responses with a meaningful organizational unit, coding of the individual responses by department is common practice.

To be useful, the survey must be conducted periodically. The principal value derives from indications of change from a previous survey. This implies that there must be some consistency in both content and format. Refinement should be incorporated in successive surveys, but they should not sacrifice the basis for comparison.

In medium and larger organizations, such surveys provide a most useful indication of "relative performance". If in response to the question "How do you rate the accessibility of your manager?" a particular unit departs from the norm, this may be a noteworthy indicator.

The word "indicator" should be emphasized here. What is surveyed are opinions, some of which may be based upon faulty perceptions. If a significant segment of the population indeed has the wrong perception, this certainly must be addressed.

To assure serious participation in future surveys, that is, to make surveys meaningful, it is essential that management provide prompt feedback. For example: Among other things, a survey can give management an indication of the value employees place upon various fringe benefits. Assume that one such benefit, let's say, vacation time, is generally perceived to deserve higher priority. It is essential that management at every level take the earliest opportunity to inform everyone, what if anything it is planning to do about vacation time and why.

Without such feedback, employees perceive surveys as a way for the personnel department to keep busy. Participation will be either minimal or perfunctory. A first or one-off survey, particularly when motivated by a perceived "problem," is not likely to produce useful results. Until the participants understand and accept the intent of the survey, the responses may not be meaningful. But once the idea has been sold, it can be a most effective management tool. I have seen close to 100% voluntary participation on the part of professional and nonprofessional employees. This suggests that they have seen feedback and consequent results with which they identify.

There is a well-established methodology to such surveys and hence no need to reinvent that wheel. Computer networks in Hi-Tech organizations have made online surveys the preferred practice. Convenience and rapid evaluation for feedback are the principal reasons.

8.3 Compensation

In societies where creative work flourishes, remuneration plays a very significant role. It may be downplayed in the creative environment, but only as long as existing variances are generally accepted. The following is illustrative of common practice in salary administration. It is not intended to argue the particular virtue of a given approach.

The distribution of salary increases, when not dictated by union contracts or institutional regulations, is generally based on merit. The appraisal usually provides the basis for the merit system. Here is how a typical system works.

The organization first of all establishes the total funds to be allocated to salary increases for the coming fiscal year. This is most often related to the projected net income. This total amount is then allocated to the functional units on an adjusted per capita basis. The adjustment may be based upon the average salary of the unit.

The people responsible for compensation in the functional unit, generally the personnel department, must then provide the individual managers with guidelines for allocation of salary increases. There are two general approaches, with numerous variations.

8.3.1 The Hybrid "Experience – Merit" Approach

All salaries within the functional unit (along the ordinate) are plotted as a function of year of graduation (Bachelor level). A family of curves, with a gradually decreasing positive slope, defines regions within the graph which separate the high, medium, and low performers as related to their performance evaluation.

Salary increases are then determined by relating a person's position on the salary axis of the graph and the position dictated by the person's performance rating. The relative increase (in percent of current salary) is adjusted to bring each person into the region corresponding to their latest performance rating.

In the interest of maintaining the privacy of personal records, the individual manager may not actually see such a graph. What is provided is a matrix which relates relative position on the graph (Current Position) to the current Performance Level, and defines two values for each matrix position. One value is the relative salary increase in percent of current salary; the other the projected time from the last adjustment, at which the increase is to be given.

2012 Salary Plan

Performance level	Current position		
	Lowest	Medium	Highest
Highest	9%—11 Mos	7%—13 Mos	6%—15 Mos
Medium	6%—14 Mos	5%—16 Mos	–
Lowest	5%—16 Mos	–	–

This grid is accompanied by a table (see below) which relates current salary ranges to experience level and is usually restricted to the experience levels of the people for whom the manager is responsible. (The numbers here associated with experience level are codes and have no absolute significance. In some organizations, experience levels are the basis for titles such as junior engineer, associate engineer, staff engineer, advisory engineer, and senior engineer. Promotion from one level to another usually involves a small additional percentage increment in the salary adjustment.)

Monthly Salary ($)

Level of experience	Low range	Medium range	High range
68	7,400–8,500	8,501–9,600	9,601–10,700
67	6,700–7,700	7,701–8,700	8,701–9,700
66	6,050–6,958	6,959–7,858	7,859–8,758
65	5,470–6,279	6,280–7,100	7,101–7,918

For example: Jane's current salary is $7,600 per month and she is in experience level 67. This places her in the low range of current earning. Her current performance evaluation is "high." Accordingly, an increase of 9% is projected for 11 months from her last increase. If her current performance were "medium," her projected increase would call for a 6% increase in 14 months.

The grid values must be designed such that current salary distribution and expected performance distribution yield a total which does not exceed the budget for salary increases. This means that there must be some guidelines for the distribution of performance ratings; i.e., the evaluation of members of a unit cannot be skewed toward one end of the distribution. These guidelines are analogous to those used in distributing grades in undergraduate schools.

The above approach makes explicit allowance for experience by relating current salary level to year of graduation. It is losing popularity, since it can leave an organization open to charges of age discrimination. One could argue that this is not a true merit system, since older people generally earn more than younger people. On the other hand, since the slope of the salary curves generally decreases, older people receive lower percentage increases than younger people.

8.3.2 The "Merit Only" Approach

The second, fairly widespread basis for salary administration is based upon a ranking system which includes, for example, all professionals in a development laboratory. In principle, this is a straightforward implementation of an absolute merit system. Assuming that starting salaries do not change with time, all increases are directly related to a person's ranking. The most highly ranked gets the highest percentage salary increase. Given a known distribution of current salaries and the budget for the year, an appropriate range of percentage increases can be determined and assigned on a continuous or stepwise scale.

There are several complications however:

- Starting salaries are not constant with time. In a competitive Hi-Tech environment, starting salaries are often negotiated over a considerable range, particularly for experienced people.
- It is difficult to change to a system of compensation based upon merit ranking, from one based upon some other criterion. People are bound to resent any change in the rules which they perceive to affect them adversely. Some managers will resist such change until they can convince themselves that the system is workable.
- The age distribution in some organizations may be such as to make a significant portion of the population see its interests threatened.

8.3.3 Ranking Criteria

As suggested in the following discussion, the choice of an acceptable set of ranking criteria is a very demanding task.

- The criteria used for ranking are likely to be different in a service group, a development laboratory, an applied research department, and a basic research department.
- In ranking an individual, consideration must be given, not only to the tangible results, but also to the success in getting started, getting organized, building tools, or instrumentation, all of which may not have any payoff until some years later.

- A ranking system makes it necessary for each functionally distinct unit to establish and disseminate a well-defined set of criteria used as the basis for the ranking.

The notion of such a set of criteria is central to the management of professional people. It is here that management can exercise leverage in placing greater weight upon cooperative effort, for example. Failing to do so can lead to fierce internal competition which could be counterproductive.

Another important criterion which enters the ranking of creative Hi-Tech professionals is that of *technology transfer*. This essentially rewards a professional not for a good idea, but the effort expended in bringing it to bear upon the goals of the organization. In most Hi-Tech environments, selling a new idea to management or to a sponsoring organization cannot be delegated to sales personnel. If the unit depends upon such effort, it must be a criterion for reward.

Even more important is the encouragement of risk taking. Too narrowly conceived ranking systems will discourage high-risk projects which have the potential for very high payoff.

Only experience and constant attention will serve to develop a meaningful set of criteria for ranking creative people in a Hi-Tech environment. For an example of such a set of criteria which has evolved in an industrial R&D laboratory, see Appendix A.2.

The performance evaluation system, if it is to serve as a basis for compensation, must be adapted to take into account those additional factors, which are not encompassed by the results achieved in a given evaluation period, but which represent an essential investment for the future. Recruiting and the development of new managers are among the criteria for the evaluation of managers which fall into this category.

By its nature, the ranking system involves all managers at all levels. Managers must merge their lists with those of other managers at the same level. This allows for discussion and, hopefully, better understanding of the different objectives of the subunits that make up a department or division. These discussions are also useful in achieving a consistent approach throughout the organizational unit. Inconsistent administration of a ranking policy is a frequent basis for employee dissatisfaction. In reviewing disputes about a particular ranking, significant inconsistencies are readily evident to the reviewer, much as they were to the person who has lodged the complaint. This is not to ignore the fact that there are chronic complainers who will challenge any decision that, by its nature, has a subjective element.

It is not always easy to compare the performance of an individual in a supporting role with that of a person responsible for management or innovation. In the extreme, each can make the difference between success and failure of the organization.

8.3.4 Ranking: Public or Private?

One last issue relates to the system of compensation based upon ranking. Should individuals know their absolute ranking relative to other people? Should the rankings be made public?

There is no consensus on this matter and answers differ widely. One school of thought holds that a person should only be told that he/she is in "the upper quartile" or "the lower third" of the rankings. This view recognizes the fact that it is very difficult, if not impossible, to defend the relative rankings of numbers 56 and 57 out of 200 people.

The matter of whether rankings should be made public has advocates at both extremes. In American society, open competition among individuals is more readily accepted than in Europe. The salary classifications of most people in Civil Service are public. It seems that public rankings are well accepted in some of the most prestigious organizations in the U.S.A. Public ranking in competitive sports is certainly widely accepted.

Those who are against making rankings public believe in the sovereignty of privacy. They also believe that some of the people at the low end of the rankings may accept their compensation as satisfactory, without wanting to be confronted with a judgment of their relative worth. Indeed, they may be working to the best of their ability. Telling them that they are among the low performers may not serve a useful purpose.

In general, the criteria for what is and what is not an appropriate issue to be raised by the manager in the course of an evaluation or in connection with ranking, should be the manager's conviction that such a discussion will contribute to the improvement of the employee's performance. If this criterion is not met, or in doubt, a manager is likely to do more harm than good. This is not to be taken as a basis for avoiding responsibility when there is a recognized need for telling it the way it is.

The viewpoint that has been put forward here is more generally applicable to professional employees than to "support" people. In the case of most professionals (as in the case of some support personnel), the assumed level of education, intelligence, and maturity would justify the expectation that responsibility is to be shared between the manager and the person involved.

It is not surprising that earlier systems of compensation were based upon age or seniority. They were certainly more easily administered and their utter simplicity made them less controversial. With time, involvement of all levels of management in arriving at the most effective system has become an important aspect of organizational development. Nor is it surprising that the difficulties inherent in every "system" of compensation have led organizations to look for ways of managing by exception.

Small companies and particularly new ventures often attract personnel by financial incentives other than salary, such as stock or stock options. This is often done out of necessity, i.e., the need to conserve a limited cash position. It puts the employee in a position of sharing in the potential gain as well as risk of the new enterprise. The gain may be realized if and when the company goes public or is acquired. The risk is a relatively low salary with relatively low job security.

The fact that the value of innovative ideas usually cannot be determined until the new idea, product, or service has been accepted by the public means that special awards are often used to recognize such achievements. It is important that the organization has well-defined criteria for establishing authorship, magnitude of contribution, and the size of the award. There are often more people convinced of the significance of their contribution than the number of awards given. In the German chemical industry, individual inventors receive extra compensation related to the revenue associated with patented products and thus are occasionally among the highest paid individuals in major concerns. Variations on this idea are found throughout industry.

Many companies have incentive programs for their executives. People in marketing and sales generally have a commission plan, which accounts for all or a portion of their compensation. Such compensation plans have many variations and are beyond our scope here.

There are data which suggest that a significant number of companies offer employees profit-sharing or cash-incentive awards, outside regular merit increases, or management bonuses. The percentage of companies that offer such compensation ranges from 32% to 10%, for cash profit sharing, lump sum individual incentive, productivity bonus, and team bonus.

One of the measures of a well-managed organization is the perception of its members that the system of compensation is equitable. This perception can be upheld only by continuing attention to the changing needs created by the Hi-Tech environment.

8.3.5 Nonmonetary Compensation

In the hierarchy of needs typical of creative individuals in the Hi-Tech environment, remuneration does not appear at the top of the list.

The opportunity to do creative work, the individual's role in influencing the choice or direction of the work, resources commensurate with the task, stimulating colleagues, and the opportunity for recognition often take precedence over salary. This is not to say that many people, particularly once they have family responsibilities, are prepared to accept compensation considered far below the norm. But within 15% or 20% of the "going rate", the above factors would probably play a more important role in the "satisfaction of needs" than monetary compensation.

Near the end of the book, in an introduction to the behavioral sciences, there is reference to the concepts of a "Hierarchy of Needs" (Sect. 15.2) and "Sources of Dissatisfaction" (Sect. 15.3) as introduced by Maslow and Herzberg, respectively. Their work bears out the assertion that remuneration is not the top priority of highly motivated scientists and engineers.

It is incumbent upon all levels of management to recognize and find ways of satisfying these nonmonetary needs. In organizations which are constrained by rigid salary guidelines, nonmonetary compensation takes on even greater significance.

There is no better way to reward outstanding performance, than to include the outstanding performer among the group of people who evaluate proposals. In this fashion, whatever superior judgment has produced exceptional results can be brought to bear on the efforts of others. Likewise, a participative or consulting role in the planning process is a very meaningful way to recognize current or recent achievement (see also: Task Force Participation Sect. 16.5).

If the organizational structure is meant to nurture a creative environment rather than serve the personal interests of those who are in a position of control, there is no better way of facilitating the planning process than by including those who are currently making the most significant contributions.

The following example illustrates the possible consequence of rejecting or ignoring this concept.

A government research laboratory in Europe lost an outstanding scientist. Upon her return from a year in the U.S.A., she wanted to start a new project or move to another department in the same laboratory which already pursued such a project. The department managers could not agree upon a reallocation of resources which would make this possible. The scientist, who had many personal attachments to the laboratory, reluctantly left for a position elsewhere.

The laboratory not only lost a valuable contributor, but it also gave others the message that the control of departmental resources takes precedence over the support of creative effort.

Top management of that organization failed to recognize the unique opportunity presented by this situation. It could have met the needs of a creative individual and at the same time changed an established practice which is likely to have a continuing negative impact upon the creative effort in the laboratory.

> In the Hi-Tech environment, an organization's responsiveness to the initiative of people who have demonstrated creative ability is the most important measure of the quality of management.

"Special Awards" are a form of compensation which goes far beyond the amount of money involved. They serve to increase the self-esteem of the recipient, bring recognition from the peer group, and provide a way for the family of the recipient to share in the fruits of effort which may have taken time otherwise devoted to the family. It is also a way of letting everyone know the priorities of the organization and its emphasis upon values such as cooperation and safety.

A type of award which is especially effective in these terms comes in the form of a "dinner-for-two." Managers at all levels of the organization are authorized to make a certain number of such awards each year without further approval. These awards are intended to provide an immediate and spontaneous expression of recognition and gratitude on the part of the manager for some effort which went clearly beyond the expected.

Awards that are more formal would be based upon a set of criteria appropriate to publications or patents or various types of achievement, be they scientific, technical, or administrative. The value of recognition from awards can be increased, by prominent posting of the pictures of recipients and by special periodic functions designed to honor awardees and their families. While I have heard expressions of skepticism about the appropriateness of such awards, I have never encountered a scientist or engineer, however senior or revered, who failed to react positively to this type of recognition.

Another sort of compensation, which is meaningful to the majority of people, is praise, both private and public praise, provided it is justified and sincere. It seems nearly impossible to praise excessively. Anyone who has ever been involved with a volunteer organization has come to appreciate the importance of recognition and praise, in this case often the only compensation. Most professional societies, which play a most valuable role in science and technology, compensate their officers largely through recognition. The enormous pride a schoolchild shows in having received the "Best Student of the Week" award reflects a human need, which is often denied due to cynicism, pseudo-sophistication, or lack of sensitivity.

As in the case of monetary compensation, the system of awards must be perceived as equitable. The public nature of awards makes particular demands upon management to devise procedures that insure the fairness of the system used.

The manager responsible for the performance evaluation of an individual to be recognized by an award should always be involved in an award decision. This manager should be responsible for recommending the award, to be approved at higher level. If managers are expected to assume responsibility for the performance of the people under their supervision, they must be seen by these same people as the person representing their interests.

There is no better way to conclude this discussion than by pointing out an eminently successful approach to the recognition and reward for creative excellence.

The Grand Challenge is the name for a project selected annually from competing proposals. The selection criteria include:

- Outside of scope of current efforts
- Potential of major impact upon significant problem
- Collaboration from within and/or outside organizational unit
- Based on existing expertise

In supporting such a project, the organization demonstrates the value it places upon striking out in new directions when ideas of truly major potential impact surface, even outside the scope of their current objectives.

It channels the energy devoted to under-the-table projects into a more integrated pursuit, it puts a premium on cooperative effort, it provides recognition for thinking outside of the box, and it builds on existing expertise. It affords high profile recognition of the value of creative effort and the importance which the organization attaches to it.

The people who have conceived the project not only get recognition but the supreme reward of an opportunity to pursue something dear to them, with high risk hence not subject to conventional ROI consideration.

8.4 Management Style

8.4.1 Is There "A Way" of Managing People?

There are several reasons why there is no *one* way to manage people.

1. There are differences in organizational personnel policies which will be influenced by many factors. Among these factors are the type of industry, the size, and maturity of the organization, current economic conditions, geographically defined practices, and the presence or absence of unions. Universities and government research laboratories have a unique set of benefits and constraints which define policy.
2. Management involves people—managers and subordinates. Every manager must answer to someone. The president must answer to a board of directors. The board must answer to the stockholders. At times, people are unreasonable. You can't resign every time you have a difference of opinion with whomever you must answer to.

Given that different people react differently to a given situation, you cannot assume that there is one way of dealing with all. Rather, if one approach does not appear to work, try another. Keep your attention on the objective, on the desired results—not on the method of getting there. This is not intended to imply that the desired end justifies inappropriate means; it is merely a plea to maintain flexibility in approach.

An extreme example: Stan comes to his manager's office in a state of agitation.

He accuses him of being a lying son-of-a-bitch.

You can certainly imagine a variety of thoughts that might go through the manager's head:

- Stan easily gets excited and is given to extreme reaction.
- I am aware of the particular incident which caused Stan's reaction and will calm him down.
- I don't know why Stan is upset. This is not like him.
- No one talks to me that way, ever!
- Is Stan right? Do others share his opinion?

This last item is probably not the first reaction of the manager.

Should it be?

There are two issues for the manager to resolve:

1. What caused Stan to react in such an extreme way?
2. What is the manager going to do about it?

The latter issue surely depends upon the resolution of the first, and the options available to the manager. Given the situation, the manager may not learn the cause of this incident.

Seeing that both people are not likely to be at their rational best, the manager may suggest that under the circumstances it may be better to try to resolve this matter the following day and get up and leave the room.

The manager may well feel like firing Stan on the spot. Is that decision likely to prevail if challenged?

To tell Stan that you find his approach offensive does little toward resolution of this situation. Stan intended it to be offensive. Once the matter has been resolved, it may be appropriate to point out to Stan that you do not appreciate his approach.

No amount of management training or playacting prepares one for such an occasion.

Another example: Dan, one of the most competent and stable managers I have known, had for years presided over weekly staff meetings of the managers reporting to him. I had never seen Dan react to anyone in an irrational manner, given a wide variety of people and situations, some of them difficult indeed.

This changed overnight when Bruce joined this group of managers. Bruce succeeded in upsetting Dan on a weekly basis by pseudo-humorous attempts to ridicule Dan's opinions or decisions.

Dan succeeded in not letting this interfere with his effectiveness, but it caused him so much personal anguish that he had to move Bruce, an otherwise competent person, to a position reporting to someone else.

This example is intended to illustrate that even a management style, which has served one well over a considerable period of time, cannot be counted upon in every situation. Each of us must accept our strengths and weaknesses in dealing with people.

Managers who stick to or defend a single "best" approach to dealing with a problem most commonly fail in the area of communication. There may well be a way of formulating assignments which work best for the majority of employees. If, however, it does not work for a particular, potentially valuable employee, it is up to the manager to try an alternative approach. This may not be the most efficient way for the manager to communicate, but it may be the only effective way.

As we mentioned in connection with the appraisal process, some employees like to have you "lay it on the line," while others fall apart when confronted with personal shortcomings. When in doubt, you may approach a problem with the opening remark: "I may be totally wrong in my perception of this matter, and please tell me if you think this is the case, but here is the way I see..." The guiding principle should be the manager's desire to help subordinates attain their performance potential and thus maximize the contribution to the organization.

8.4.2 Guidelines Versus Rules

Standards of performance or behavior, however defined, should be understood as guidelines. A manager is expected to follow these guidelines—but not blindly. If there is a good reason to depart from the guidelines, the manager should be prepared to defend a specific action in a rational way.

To label a rule as being "a stupid rule" is not a rational defense of managerial action which departs from the rule. But to point out the reasons why in this particular case compliance would not produce the desired results may be the basis for reviewing the general applicability of a given rule or the circumstances which have changed since the rule was conceived.

Only by acceptance of the notion that "management is us, not them" can management guidelines be changed by those responsible for current implementation. All rules are made by people and only people can change them.

Not only is there no *one* way to manage, there is always a *better* way. As managers, we are expected to find it and to change the rules accordingly.

8.4.3 Management Style: Perceived Positively

The management style which I have come to admire most is that of the person perceived as low key and easy going. This person is not only a good listener but makes you feel that your concerns are of the utmost importance. (The manager reading his mail while you are talking does not project this image.) The positively perceived manager is decisive in terminating a meeting and otherwise acting in an assertive manner, but only when the occasion demands. He or she has no difficulties with admitting error or accepting the suggestions of others and no need to demonstrate superior ability. (A manager's perseveration of superior exploits on the golf or tennis court belies this image. Anecdotes about heroic deeds of the past have a particularly negative effect.) An altogether positive characteristic is sincerity in praising others and an inclination to do so at every appropriate opportunity.

Above all, the positively perceived manager does not dominate the discussion, but is always ready to make a decision when the occasion demands.

It is remarkable how many executives of successful and well-managed organizations fit this mold. Admittedly, there are glaring exceptions (Wallace 1992).

8.4.4 Management Style: Perceived Negatively

There are certain traits and characteristics which are not uncommon among managers with a science or engineering background which I find undesirable.

- The need to demonstrate in-depth understanding of the technical issues, beyond what is required to make a decision. When this is more evident in situations involving the area of the manager's expertise, it betrays insecurity in other areas of decision making.

* Deliberately idiosyncratic behavior which conveys arrogance by flaunting the ability of getting away with it. This is almost certain to undermine the confidence of the people who feel dependent upon the manager—superiors, colleagues, and subordinates alike.
* The multiprocessing executive who carries on several activities simultaneously. Since most people cannot do this effectively, this manager is perceived as unable to cope with the job or trying to impress you with a virtuoso performance.
* Unacceptable in a manager is the need to put people down. This is a trait which is incompatible with my idea of a manager and, for that matter, with my idea of people I want to associate with. Frederick Herzberg, the social psychologist, has found this trait one of the most destructive in human interactions: "... perhaps the most ubiquitous and damaging is that of degrading others so that it is possible to look better in comparison" (Herzberg 1966).

The value placed upon effective managers in the world of Hi-Tech is so high that tolerance for such undesirable traits is not unusual. Our subsequent discussion of organizational culture (Chap. 13) puts this problem in perspective.

8.4.5   Summary: Managing People

The subject of recruiting, from the viewpoint of the applicant and of the recruiter, appropriately takes first place. If the basis for selection and the associated process is inadequate, it becomes a problem of managing the wrong people.

The importance of performance evaluation cannot be overstated. It is an activity for which scientists and engineers are generally not prepared—either intellectually or emotionally. Without this most valuable task-relevant feedback, it is difficult for people to realize their creative potential. The procedures suggested here are guidelines to be adapted to each specific situation. The effectiveness of performance evaluation, in terms of benefit to the individual and to the organization, has been firmly established in the world of Hi-Tech.

Creative professionals generally do not place compensation among their prime concerns. This is based on the assumption that an equitable basis for compensation exists. We are providing a rationale for that assumption by pointing out some of the approaches and associated issues.

The most subjective aspect of people management is that of management style. This refers to the way managers are perceived, correctly or incorrectly. Our discussion should heighten awareness and help readers to formulate their own ideas on the subject.

We have expressed some strong views on the subject of people management, with the hope of provoking the reader's reaction to the issues discussed. There is *no one way* of managing people, but always a *better way*.

9. Evaluation of Research and Development

Unique to the Hi-Tech environment is the level of technical knowledge managers must bring to the evaluation of R&D projects. What considerations, other than technical, are relevant to the evaluation of ongoing and proposed projects?

In the following discussion, the evaluation of R&D activities by levels of control is used to indicate the distribution of responsibility among the various levels of management. *Project Level* is here defined as the lowest level of management. The *Department Level* implies an aggregation of several projects. The *Top Level implies* ultimate responsibility for the organizational or strategic unit. When an organization is large enough to justify additional levels of management, the responsibility for an intermediate-level evaluation should be associated with the authority for commensurate resource reallocation. More than three levels of evaluation become burdensome and are to be avoided.

There is a considerable variation in the frequency of evaluation at various levels and in the degree to which such evaluations are formalized. The relative importance of the project(s) and the benefit to the evaluated should be the principal criteria for frequency and scope.

9.1 Evaluation at the Project Level

The project manager, of course, constantly evaluates the project to initiate such changes as may be necessary to meet the given objectives and schedules. When the project in question is part of a larger coordinated effort, the project manager must also maintain an awareness of the activities of other groups that depend upon the results of the project as well as other projects upon which the manager is dependent. Hence, objectives and schedules must always be considered within the larger perspective of the relevant organizational unit.

Most projects involve the cooperative efforts of several individuals. Weekly meetings can serve to review progress, exchange ideas, and learn about current concerns and potential problems. Among such concerns are personnel problems

R. Kay, *Managing Creativity in Science and Hi-Tech*,
DOI 10.1007/978-3-642-24635-7_9, © Springer-Verlag Berlin Heidelberg 2012

which might impact the project, scheduled or unscheduled absences as well as potential bottlenecks which may be anticipated in a "what if" situation, etc. Regularly scheduled project meetings should not only help identify potential technical problems, but also promote a common understanding of such problems among the members of the project. Resolutions are likely to come from anyone directly affected by the success of the project. The meetings also serve to prepare the project leader, and for that matter all members of the project, to respond to requests for formal presentations of the project to those with a need to know.

It should be understood that the value of most projects prior to their completion resides in what is communicated about their progress and promise to those who control the resources. Since such presentations often require schedule and budget status and critical dependencies, this information must be shared in the project meetings. What is more, it is a source of personal embarrassment for a project member to have to turn down an impromptu request for a project presentation for lack of adequate knowledge of the current status.

Evaluation at the lowest level must identify and communicate potential problems or important breakthroughs in a timely fashion. Without the necessary in-depth understanding of the implications of unexpected and significant issues, higher levels of management are likely to overreact. This can mean loss of control and the much valued independence of creative people.

If problems are recognized at the lowest level, it is incumbent upon the project manager to ask for review at the next level without delay. Nothing undermines the confidence of higher management more than the discovery of unexpected problems.

9.2   Evaluation at the Department Level

A broader perspective than that of the project level is necessary for redistribution of resources between projects and to consider major changes which would lead to the termination or redirection of a project. It is assumed implicitly that such changes imply a more pressing or attractive alternative in terms of objectives, potential impact, or success potential. If such alternatives are not readily available, it may be prudent to devote some or all of the resources of a failing project to the definition of a more attractive alternative. In an earlier discussion of the desirability of managing creative people, an explicit example of such a situation was considered (Sect. 3.5).

At the intermediate level of management, there is often the greatest reluctance to accept the fact that a project does not live up to expectations, and that the addition of time and/or resources will not make a significant difference in the ultimate outcome. On the other hand, there is also the tendency, in an environment that has a significant creative component, to premature abandonment of an idea or approach for what appears to be a better one. Only experience relevant to the organizational unit can be counted upon to identify issues of this nature. Most of that experience, presumably, resides at the top level of management which is also less likely to be encumbered by parochial interests.

Evaluation by an intermediate level of management also provides an excellent vehicle for organizational development. Project leaders and project personnel cannot gain adequate experience from the necessarily infrequent review by top-level management.

9.3 Evaluation at the Top Level

Periodically, those responsible for the overall success of an organizational unit must review current efforts in order to evaluate whether the present distribution of resources is optimal in serving the existing or anticipated strategic direction. Embryonic efforts with the potential for new strategic directions may warrant special attention at this level.

At issue here is the extent to which the strategy is market, technology, or finance driven. In order for any Hi-Tech product or service to find acceptance, there must be an existing need or the product itself must be capable of generating such a need.

A *technology-driven strategy* may aim to develop:

- A new product or service made possible by new technology. Bioengineering-related products and services are a current example. Man-made fibers have produced a plethora of textile products which had not been considered before.
- A new market for an existing product or service made possible by better cost, performance, and/or function based on new technology. Desktop printing has brought composition and layout to the person creating the document, by virtue of improved cost, performance, and function and thus created a new market.
- Smart phones were made possible by reduced component cost, size, and power consumption, thus providing increased functionality.

A *market-driven strategy* will seek to:

- Protect and enlarge a market, by making sure the needs of the existing or sought after customer base are satisfied. It may depend heavily upon new technology to realize this objective.
- Relative market share has other implications. A company that controls 50% of a billion-dollar market can spend a great deal more to achieve a marginal cost reduction than a competitor who has a few percent of the market. On the other hand, with a given investment, the one with the small market share may gain more from advances which increase its market share.

Financial opportunities or needs may dictate strategy at a given point in time. The financial considerations which would motivate acquisitions or mergers are not unique to the Hi-Tech environment (Riggs 1983).

What has gained relevance in the world of Hi-Tech is a point made by Ralph Gomery in his Scientist-of-the-Year lecture. He draws the attention of top-level management to the need to recognize that it is *the speed of the development and of the manufacturing cycle* that appears as technical innovation or leadership.

That is to say, the potential of a new idea, in terms of effective technology transfer, is constrained by its window of opportunity (Gomory 1987).

At the highest level of evaluation, it is essential that there exist the insight to *judge the timeliness of an idea*. This judgment is reflected in the consequent allocation of resources which will determine the rate of progress to be expected.

Strategic considerations cannot enter the evaluation without concern about the operational implications.

Example: This large project aims to develop software designed to work on a particular type of computer. Many of the firm's customers use this type of computer. The project manager becomes aware of the impending availability of new, more cost-effective computer hardware.

Should the currently targeted hardware be abandoned to take advantage of the promise implied by the new hardware? Among the relevant questions are the following:

- What commitments have been made to current hardware?
- Who will be affected if these commitments are not met?
- Will key personnel resign to work on new hardware?
- Are the product and/or market implications for the new approach consistent with related strategic considerations?

The current project may be based upon the needs of a large customer base working with existing hardware, while the new hardware will not equal this installed base for at least 5 years, if ever.

The project is 9 months from completion; if several key people leave, completion may be delayed for a year or more, at which time competitive products may have made inroads.

Several other programming projects are dependent upon the availability of the program under development.

This example was designed to illustrate the potential conflict between strategic and operational considerations. If the key people in the project can be involved in the decision, the feasibility of continuing with the current approach can be evaluated. They may be persuaded to complete the project, with a few of them working part-time to define a follow-on project to take advantage of the new hardware.

The example also points out the value of a "Participative Group" style of management as defined by Likert (1967) and discussed in Sect. 15.4.

It should be understood that strategies have no virtue in and by themselves. They only serve to focus upon primary goals of an organization at a given point in time.

Evaluation of projects must take place within the framework of an existing strategy, even though the evaluation can lead to a change of some elements of the strategy.

9.4   The Importance of Under-the-Table Projects

One of the most important elements of a strategy in an organization with a strong R&D component is the continuing emphasis upon new, exploratory efforts.

Such efforts often take the form of "under-the-table" or bootleg projects.[1]

The formal planning of R&D efforts must leave sufficient slack to allow for such activities.

In some organizations this slack is obtained through the creation of "special studies" projects which provide a temporary home for people engaged in exploratory efforts. Slack may also be created by allocating less than 100% of project resources.

Some of the most successful Hi-Tech organizations have allocated a percentage of an individual's time to projects of their own choosing (Levy 2011). Another very successful way to institutionalize such exploratory efforts is the annual competition for "The Grand Challenge" project discussed in Sect. 8.3.5.

Evaluation of R&D efforts which aims at an accounting of man-hours is likely to stifle creativity or lead to dubious accounting practices. To find the right balance between a tightly run project and adequate slack for new creative effort is one of the more difficult management tasks. It is very much a function of the people involved and the nature of the project. Attempts to find "generally applicable" guidelines should be resisted.

The best approach aims to delegate the responsibility for providing adequate slack time as far down as possible. Ideally, creative individuals should be able to decide how best to allocate their own time in doing justice to the planned objectives and the need to pursue longer term alternatives.

It is also important that managers who find ways to support such relatively high risk projects are given recognition for taking risk, independent of the success of the effort.

9.5   General Observations About Project Evaluation

Experience suggests that it is very difficult to terminate a once-promising project. Over the course of time, a lot of people will have a vested interest in the success of the project. After all, continued support was based upon their best judgment. For this reason, evaluation in the early stages of a project is more likely to be effective. It seems that the larger the project and the longer it has been active, the more difficult it is to terminate. Ideally, a project starts on a small scale and periodic evaluation serves to justify incremental increase of resources. Additional resources can be a motivation to reach intermediate goals.

---

[1] The German term for such projects translates to "Submarine Projects": they only surface when in need of provisions.

There is nothing more frustrating and costly than the need to terminate a grandly conceived big project, the goals of which have been recognized as unattainable.

Unexpected problems or reevaluation of known problems, technical or managerial, are generally the focus of evaluation. Well-defined milestones help to distinguish between various kinds of problems: Problems which are beyond the control of the people involved and those problems which reflect directly upon project management and personnel.

For example, delay due to a change of specifications over which the project had no control points to very different management action than a schedule missed due to the loss of key people, suggesting a possible personnel problem.

Repeated failure to meet schedules and budget commitments often point to something being amiss in the management chain.

- Was the schedule imposed without adequate consultation with the project manager?
- Did the project manager fail to make project personnel aware of commitments?
- Does the project manager consistently underestimate or overcommit?
- Does the project manager appreciate that a significant part of the organization can judge him/her only relative to demonstrated ability to meet commitments?

There is something like an organizational time constant which defines a point in time when a project should be facilitated or terminated. Not that this time constant can be defined in rigorous terms. If the project continues to hobble along over an extended period of time, project personnel may perceive this to be a lack of endorsement · and lose motivation.

9.6 The Evaluation Process

The success of each manager, regardless of level, depends to no small measure upon an ability to facilitate the evaluation process. The manager's presentation does this by concentrating on those aspects of the project which are relevant to the decisions likely to result from the evaluation, primarily, an increase or decrease in resources or a change in the schedule. Hence, the question as to the potential impact of increased or decreased resources must be anticipated.

History and distant expectations, or tutorial embellishments, rarely serve the purpose at hand. Nor is this the time and place for the manager to point out his/her own technical contributions to the project.

The evaluating manager(s) can contribute to the success of the review by giving the less experienced project manager some prior indication of what is judged crucial to the evaluation. The evaluating manager(s) must also be sensitive to the personal reaction of the people whose project is being evaluated. Destroying the self-esteem of the presenter can rarely be justified by the evaluator, be it through the desire "to get to the bottom of things" or the need to demonstrate superior insight.

An overly negative or critical attitude (on the part of evaluators) toward ideas in the incubation stage may destroy the motivation to pursue an important problem. A consistently negative or "hard-nosed" attitude can reduce the readiness for risk-taking which is an important element of creative effort. The early attempts at a solution can provide the impetus for subsequent refinement. Or to paraphrase the late John Backus, a pioneer in the field of computer science: "The failed solutions which make up the junkyard of my ideas provide the tools used to build the systems of the future" (Backus 1984).

In the R&D environment, new ideas are the most precious of all commodities. Management must create an environment which encourages new ideas, nurtures the ones which prove to be relevant to the issues at hand, and recognizes others as potential enrichment of the tool kit. The evaluation process is a good vehicle for establishing such an environment.

Early on in a project, the most relevant questions on the part of the evaluator address the matter of alternative approaches and potential impact.

- What are the alternative approaches you have considered?
- How would you define "success" upon conclusion of your project?

The payoff for a project often materializes only upon successful technology transfer. It is therefore important that progress in the planning and in the commitment to technology transfer by the intended receiver be part of the ongoing project evaluation.

To summarize the criteria for evaluating ongoing research and development projects:

- Has the success potential of the project increased since the last reviews?
- Has the project met schedule and budget objectives? If not, why not?
- Assuming successful completion of the project, has its expected impact increased?
- Have external conditions enhanced the success potential and potential impact?
- Has the potential for successful technology transfer increased?
- Does the project serve the strategic objectives of the organizational unit?
- Does the project provide an opportunity for new strategic initiatives?
- Does the project enhance the competitive posture of the organization?
- Does the project enhance the knowledge base of the organization?
- Does the project contribute positively to the morale of the organization?
- Is there common understanding among all concerned about current and anticipated problems and the priorities assigned to their solution? This is one of the most important aspects of an evaluation.

The relative weights attached to these criteria will of course depend upon the specific circumstances defined by the environment and strategy which provide the frame of reference for the evaluation. The above order is not meant to imply relative weighing factors, but it should be understood that high morale in a failing project has little influence upon a decision to terminate the project. Likewise, the desire to upgrade the skills of an organization may be achieved more readily with a promising project, than with a project which is unlikely to survive.

To undertake an evaluation without an explicit set of criteria can be frustrating to both the evaluator and the evaluated. In the course of the evaluation the criteria may be found wanting and hence in need of reassessment.

Creative people particularly resent perfunctory reviews by people who lack the relevant ability to evaluate a given effort. Management should protect creative effort from such sources of demotivation.

It should have become clear in the above discussion of project evaluation that it constitutes a most important management tool.

> Evaluation should always be perceived as a basis for decision making.

Evaluation loses much of its value if it is perceived as education or general information gathering. One way to convey the objective of decision making to those evaluated is by prompt feedback of decisions which have been influenced by the evaluation. Even approval and the consequent decision to continue as is should be stated explicitly, with some specific reasons for the decision.

9.7 Evaluation of Proposals

We are left with the problem of judging the potential of creative effort at the proposal stage. This is intrinsically more difficult than the evaluation of an ongoing project. The proposed ideas may be largely untested. Rejection of a proposal disallows the opportunity to gain additional insight.

There are but few general guidelines available. Probably most important is the assessment of potential results. Nobel Laureate Medawar puts it most succinctly: "To achieve significant results, you must work on significant problems" (Medawar 1979).

Nobel prizes recognize seminal results. In every field of science and technology, there are significant goals which warrant above average risk.

Evaluation of a proposal is never an isolated decision. The required resources must come from existing activities or justify new resources in competition with projects that have already reached a level of maturity that reduces the associated risk.

Risk must be related to the expected results. An incremental improvement or extension of a well-established idea, be it a theory, a method, or a product, cannot justify the investment of significant resources unless the resultant benefit can be quantified. Extrapolation of the past rate of progress provides the scale; i.e., advances in science and technology are usually plotted against time.

The desire to improve a design or an experiment, or to abandon the chosen approach for a better one, is fairly common among engineers and scientists. For example: A given approach has not yielded a promising solution within the projected period of time. To the people most deeply involved, the new approach is *sure* to produce more spectacular results based upon the knowledge gained so far.

It is important to determine whether, this, let us call it "parochial attitude," is based upon the desire to continue with the familiar goal, or a tendency toward perfectionism or a valuable new insight which will be decisive in reaching a worthwhile objective.

While breakthroughs cannot be scheduled, it is nonetheless possible to apply meaningful criteria to the pursuit of new ideas. In our earlier discussion (Management of a Department or Small Enterprise, Chap. 7) we defined elements of a strategy, which provide criteria relevant to the evaluation of a proposal for a new project.

The strategy of established organizations reflects the fact that they are constrained by one or more of the required resources, be it funds, people, space, capital equipment, or executive attention. These constraints provide boundary conditions for the evaluation of proposals.

One of the attractions of the entrepreneurial start-up is the possibility of making various trade-offs among these constraints that are not possible in an established organization.

Another attraction of university research and Hi-Tech venture capital enterprise is based upon the availability of relatively unconstrained resources such as graduate students or people trained by others, and uncommitted funds. Nor are existing market share or customer needs constraining factors. Much can be learned from the approach of the venture capital community to the evaluation of proposals. Particularly noteworthy is the access to evaluators with uniquely relevant experience; see Sect. 12.1.6.

There are arguments for both, starting small or starting big. It is sometimes difficult to get much attention with a truly new idea involving significant risk when the scope of the proposal is so modest as to cast doubt upon the conviction of the proposing party. At the outset, the proposer should be explicit in providing the most optimistic estimate of the potential impact of the proposed idea, in order to suggest the level of priority the proposal should have in terms of management attention. An organization that has a plethora of proposals may prefer to start several small projects. This should increase the number of potential winners. It may also be a way of filtering small exploratory efforts by establishing feasibility before scaling up the effort. By implication, criteria and means for terminating the less-promising efforts must be available.

Certainly short-term versus long-term opportunity and risk must be weighed in the evaluation of new project proposals. Here are two examples:

Given the alternative of working on a new material which does not contain the polluting ingredient or a method of controlling or reducing the polluting effect, I would favor the new material, particularly if there were a potentially growing demand for the material. The "method of control" alternative may have greater short-term potential, but a competitor with the new material could be a real threat.

The development of two compilers, one, quick-and-dirty for the development effort, and another, "fully implemented" for the product offering, has tempted many groups embarking upon major software projects. The decisive issue is usually the availability of qualified people and their willingness to get involved with the task considered less desirable. Major efforts involving some of the most competent people in the field have floundered because of inadequate appreciation of the short-term and the long-term effects of delays due to marginal staffing.

These two examples were chosen to illustrate the range of considerations and the level of complexity inherent in the evaluation of new projects. All too often, these decisions do not get the attention they deserve. Few organizations, be it in the academic or industrial world, have developed adequate mechanisms for dealing with this topic.

By comparison, the review of ongoing projects is straightforward. With each successive review, the evaluators become more confident in their knowledge of what is relevant to ultimate success. They can judge the match between the problem to be solved and the resources required and available to solve it, based on actual, even if limited, performance.

In judging the worthiness of a proposed new effort, only relevant past experience is likely to be useful. In the case of a proposal which is orthogonal to that experience, skepticism toward the evaluator may be warranted. Even so, consistent past success based upon a particular approach can unduly influence some of the most experienced evaluators in considering radical innovation. Such innovation may be threatening the very power their past success has bestowed upon them.

It is very important that the evaluators of new projects include individuals who have a stake in the outcome, such as their reputation for identifying winners, but not be affected by the potential redistribution of resources. Expert consultants, divorced from the intramural interests of the particular organizational unit, can provide a unique service in focusing upon the technical aspects of the proposal. In dealing with untried ideas, their experience can complement that of the people who have a stake in the outcome of the evaluation.

The evaluation of proposals for new projects is one management responsibility where the relevant experience, breadth, and intellectual integrity of the person with ultimate responsibility play a decisive role. Breadth is here to be understood as the ability to bring balanced judgment to a multifaceted problem. Intellectual integrity here is the ability to resist the temptation of substituting "intuition" based upon marginally relevant experience for technical understanding of the matter at hand; see also Chaps. 5 and 11.

I do not believe in the checklist approach to the evaluation of proposals. But here are the types of question which have proven helpful in evaluating new project proposals. The questions listed below are not meant to be exhaustive, nor do all of the questions apply to every situation. The level of management involved in the evaluation would help determine the appropriateness of the question. The questions also point to some of the preparations which will facilitate a decision.

9.7.1   Evaluation Criteria for a New Project

1. *How Important Is This Project to Those Proposing It?*

   - Has the proposer invested some of his/her own time in the preparation of the proposal? This question is analogous to the question of the Venture Capitalist: What portion of personal resources is the proposer prepared to invest in the venture? (See Chap. 12)
   - Is the proposed work an extension or supplement of ongoing work, or
   - Are there people, other than the proposer who are prepared to drop what they are doing to join this effort? Who are they?
   - Have the appropriate managers indicated their willingness to commit existing resources to this effort? What are they willing to drop in order to pursue this new effort?
   - Will this project allow the redirection of existing resources currently committed to a project that is to be terminated?

2. *What Is the Potential Impact upon the Organization?*

   - Assuming realization of the most optimistic expectations, what will be the ultimate cost and the ultimate benefit? How is benefit to be measured?
   - Does it have the potential to change the direction of 0.1, 1, 5, 10, 20, or 50% of all current activity?
   - Is there a clear concept of technology transfer? To whom are the results to be transferred? Is there interest and/or commitment on their part?
   - Will realization of the benefits require extensive restaffing of other parts of the organization? Are new skills required in multiple functions? What skills, and in what functions?

3. *What Organizational Objective Does This Proposal Address?*

   - Does it promise competitive advantage or does it meet a recognized competitive threat?
   - Does it protect or extend market share, create new markets, or improve profitability?
   - Does it fill a recognized need within the organization or does it provide the opportunity for entirely new efforts?
   - Does it have the potential for a position of leadership?
   - Will it lead to an upgrading of the organization's skill level?

4. *What Are the Knowns and Unknowns About This Project?*

   - How would you rate the probability of success? What is the basis for this rating?
   - What are the critical issues upon which success hinges?
   - Is this the only way to address the perceived problem? What are the possible alternative approaches?
   - Would a smaller initial effort be meaningful?

- With whom has this idea been discussed? What was their reaction? What was the basis for skepticism, if any?
- Where might additional help be obtained if needed?
- Are there any potential patent or legal implications?
- Is authorship within our organization clearly established?
- Will we have the necessary skills, facilities, and space for early exploitation? What are the potential bottlenecks?

5. *What Are the Implications of NOT Pursuing This Project?*

- Is the proposer likely to pursue this idea elsewhere, taking key people along?
- What would the successful pursuit of this idea do for our competitors?
- Would it imply a lost opportunity of acquiring new skills?

Many of the constraints addressed by some of these questions are specific to a given organization. Some of the answers will be dictated by such constraints *and* by the outlook of top management.

9.8   Evaluation of the R&D Function

One of the most difficult and pressing issues facing many organizations, and even nations, is to find answers to such questions as:

- Is there a meaningful way of evaluating R&D?
- What are we getting for our investment in R&D?
- What is the appropriate amount of R&D?
- What are the criteria for setting priorities?

In reading about this very real issue, it is surprising to find:

- A relatively small body of literature, mostly by people in the academic world, i.e., very little from practicing R&D managers.
- Little evidence of the flow of people between R&D laboratories and academia, the kind of flow which has made the academic influence in other areas of management such a vital force.
- The Harvard Business Review, a leading journal of management, has offered little on the subject of R&D management in the past 25 years.
- Peter Drucker, widely regarded as the leading figure in the field of management science, ended an article on the subject: "Best R&D is business-driven" with the statement: "But no one today—and surely neither the engineering nor the business schools—knows how to teach technology management nor, indeed, even where to start" (Drucker 1988).

In the face of this, it may be presumptuous to offer definitive answers. Nevertheless, an effort to come to grips with these questions seems in order. Several approaches have proven effective and some useful generalizations have emerged.

9.8.1 How to Approach the Evaluation of an R&D Function

Experience has convinced me that a pragmatic approach to the subject of evaluating R&D functions has the greatest promise at this time. Pragmatic, in the sense of seeking an effective approach to solve a limited problem well, learn from it and try to extend its applicability, modifying the approach along the way. The complexity of the problem space and the need for both general management experience and more than superficial understanding of the problems unique to a particular R&D environment support this contention. The application of generally proven management principles is necessary but not sufficient. The pursuit of comprehensive solutions based on solid theoretical foundations is more likely to put useful solutions into the ever more distant future.

One may start with a set of five questions analogous to those suggested in connection with the evaluation of new project proposals (Sect. 9.7).

- How important is R&D to the people responsible for the overall success of the entity to be served by the R&D function?
- What is the potential impact upon that entity?
- What objective(s) of the sponsoring entity does this R&D function address?
- What are the knowns and unknowns?
- What are the implications of eliminating or significantly changing the size of this function?

(In considering these questions, it may be well to have in mind a problem which can be grasped in its totality, a part of the R&D function which addresses an area with which you are familiar.)

1. *How Important Is R&D?*

  - What is the motivation for R&D of those responsible for the success of the overall enterprise?
  - Is the R&D function represented at the top level of management?
  - Does top management personally identify with a need for R&D or is it accepted as a necessary expense?
  - Is the role of R&D expected to grow or diminish?
  - How does top management expect to measure R&D performance? How does R&D function expect to be measured?
  - How is the potential value of the R&D function quantified?
  - Is there a clear understanding of how R&D is to interface with other parts of the organization? (Technology transfer).

2. *What Is the Potential Impact of R&D?*

  - Ability to attract top talent? (e.g., in a university)
  - Advance or maintain competitive position?
  - Minimize risk of missed opportunity? (e.g., national policy)
  - Effect on profitability?
  - Potential for leadership in any or all areas of activity?

3. *What Current Objective(s) of the Sponsoring Entity Does This R&D Function Address?*

- Provide basis for growth?
- Provide options for new directions?
- Maintain or achieve position of leadership?
- Increase profitability?
- Upgrade or broaden general skill level?
- Fulfill social responsibility?
- Create favorable image with investors?

The word "current" was added to question 3 above, as an afterthought. It was triggered by the experience of evaluating an R&D function on behalf of a newly appointed corporate president. Before starting this task, the president provided a clear view of the importance he attached to R&D activities and his expectations and objectives. In looking at the major directions and the nature of the ongoing projects, it became evident that the rationale for their existence, by and large, had outlived their relevance to the current and future objectives of the organization. How could this have happened? The R&D function had been allowed to become isolated from the concerns of top management.

This provides the basis for the first generalization:

There is little to be gained from an evaluation of the success potential of an activity which is judged irrelevant to current or future organizational objectives.

4. *What Are the Knowns and Unknowns?*

- Is top management enabled to stay informed of progress and new developments?
- Is there an up-to-date skills inventory of the function? How does it compare to the skills distribution of the organization as a whole?
- Is there an adequate means for maintaining awareness of competitive developments?
- Is there sensitivity to the weak signals that indicate future problems?
- Is there access to independent expert opinion? Has the effectiveness of that opinion been tested?
- What is the role (and motivation) of the person initiating the evaluation?
- Is there a previously agreed upon set of evaluation criteria?

Another useful generalization:

An evaluation must be based upon a set of evaluation criteria, clearly understood and agreed upon by the evaluator and the evaluated. Such understanding and agreement must be reached at the time an activity is initiated, not just before or during an evaluation.

5. *What Are the Implications of Eliminating or Significantly Changing the Size of This Function?*

- Is this possibility currently being considered?
- What organizational objectives would be most seriously affected?
- What functions outside of R&D would be most severely impacted?

Given the risk and uncertainty associated with much of R&D, it is important to understand the motivation for the evaluation and willingness on the part of management to consider radical change. For an evaluation to be meaningful, a great deal of effort is required. The burden this imposes upon R&D personnel should not be underestimated.

Useful generalization:

The evaluator must have a clear understanding of the motivation for the evaluation on the part of those initiating it.

It should be evident that, to be meaningful, the evaluation must be based upon well-informed answers to such a set of questions appropriately tailored to the specific situation. An explicit strategy for the R&D function, such as we have discussed in Sect. 7.2, will address a good many of these questions.

9.8.2 What Are We Getting for Our Investment?

In the course of a review of an R&D function of a major Hi-Tech corporation, this question was posed to the manager of the function.

This particular R&D function had earned the reputation of having done important and original work which had subsequently been exploited more effectively by competitors; the person asking the question made reference to this fact. The response provided a number of examples of successful technology transfer within the corporation. It also cited the associated current annual revenues.

Most of the reviewers were satisfied to have a detailed answer. What the answer did not address were the following points:

- What was the R&D cost associated with the subsequent revenues?
- How did the ratio (R&D cost)/(revenue) compare to that of projects in other parts of the corporation?
- How much relative revenue growth was expected from similar investments in marketing and advertising?
- What was the relative contribution to the targeted corporate revenue growth? How does it compare to other functions?
- How much revenue did similar R&D expenditures produce for competitors in the particular product field? And for the competitors who followed their lead?

A more pointed question about corporate management's failure to exploit the results of this R&D effort as effectively as some competitors brought the surprising answer:

We had no problem with top management,—it was product engineering, manufacturing, and sales who were not up to the job.

The implication being that *"they"* (top management) should have told *"us"* (the experts), that the new technology to be introduced did not match the skills or experience of the existing workforce.

The question of what we are getting from our R&D investment must be taken seriously by R&D management. Only they are in a position to find appropriate answers.

Useful generalization:

It must be understood by R&D management that the burden of making their function relevant to the entity served is clearly on their shoulders. This means developing a meaningful way to evaluate the function, AND taking responsibility for effective technology transfer.

Ralph Gomory, at the time IBM's Director of Research, when asked about the secret of effective technology transfer, responded: "Do whatever it takes and don't stop until you have succeeded."

This straightforward maxim applies equally well to the task of ascertaining the value of R&D to the entity to be served. In large, complex entities as found in government, industry, and universities, the task can be demanding. If a single metric were adequate, little judgment would be required.

The R&D required to find a material which plays a crucial role in a critical component may be evaluated differently by the manufacturer of the component and by the supplier of the material. For example, a polymer which is crucial to the fabrication of integrated circuits may justify significant polymer R&D on the part of the circuit manufacturer but not on the part of the chemical company, which ultimately supplies this material in relatively small quantities by the standards of the chemical industry.

Some of the output of a research function can be measured by the per capita rate of publication. Beyond a simple count of publications, the citation approach has come into favor. It is facilitated by the Citation Index, a publication which tabulates current citations in a wide range of journals and thus makes possible a more meaningful assessment of the impact of a given publication. Properly used, this approach can be very useful. There are some potential pitfalls: parochial citation practices and self-citation, review articles vs. original contributions, the order in which authors' names are listed, and the variable interval between publication and citation. It is not unusual to find the leading citation count associated with the publication of widely used constants, computed or compiled by the author. Without familiarity with the particular area, it is not easy to differentiate this type of publication from a seminal contribution.

To avoid being misunderstood on the topic of quantitative measures, a general word of caution is in order. There is the view, sometimes even among top management, that everything can and must be reduced to numbers. This view is nurtured by some business schools who are into quantitative analysis as a substitute for critical thinking. In the context of evaluating creative effort, numbers are most useful when it is possible to make meaningful comparisons.

On the other hand, R&D managers who defend the notion that it is not possible to do a cost–benefit analysis of their function—that it does not make any sense—will find it increasingly difficult to get away with it. It is simply becoming less and less acceptable to those who pay the bill.

9.8.3 What Is the Appropriate Amount of R&D?

With the above caveat about reducing complex issues to a readily grasped set of numbers, an answer to this frequently posed question will be attempted.

First of all, universities and government R&D organizations do not have the same potential access as an industrial R&D function to those making or influencing budget decisions. That is to say, the alignment of interests and objectives is generally closer in the case of industry. Likewise, the ability to quantify and evaluate potential benefit is greater in industry.

Population and industry statistics are useful in understanding relevant norms. They do not provide much insight into the reasons for individual variances.

In concrete terms, the question about the appropriate amount of R&D is most meaningful when contemplating change. When the total R&D budget changes, the issue is often one of uniform allocation of the change versus making hard choices. Most administrators would take the more readily acceptable alternative to treat everyone the same. In some organizational cultures, to do otherwise would be nearly impossible. Most likely, this course will have the least impact, negative or positive.

If additional resources for R&D, beyond those needed to address recognized needs, are to be considered, it is essential to ask for proposals which address issues not addressed already. Such proposals, based upon insight unique to the R&D function, can provide top management with the basis for future opportunities and indeed provide the rationale for increasing the function's budget.

An occasional need for a significant increase in R&D resources is the sudden realization that an essential part of the infrastructure has become obsolete or inadequate in meeting the operational needs of the R&D function itself. These occasions often provide the impetus for a more appropriate way of budgeting for capital equipment (instruments, special facilities, etc.) and computing capability.

It has been difficult for me to identify with discussions about the appropriate amount of research versus development, and the appropriate balance between "basic" and "applied" research. The definition of Research and Development differs widely in different fields and even from company to company in the same industry.

If that which is labeled "Research" has been relatively more effective than "Development," first try to understand why. If the WHY cannot be realized in "Development" a relative increase in "Research" may be warranted.

The differentiation between applied and basic research seems to be of little concern to the most highly esteemed researchers I have encountered. They seem to be more concerned with being effective in whatever they do, be it their own research, teaching, or consulting. When a scientist doing basic research recognizes a unique practical application and feels personally best able to demonstrate its usefulness there is little doubt that most would look upon this as a great opportunity. This is not a matter a matter of doing "basic" versus "applied" research but a matter of maximizing one's potential contribution and reaping the commensurate reward.

To do effective basic research it is necessary to meet international standards which may be beyond reach. By these standards, there is little recognition for the second best. In some organizations, basic research is done in disciplines which intersect those of the "base technologies" which are critical to an industry. The objective is to have a highly skilled pool of people, on the leading edge of the most relevant fields of science, able to recognize significant developments early and be available as troubleshooters. Some organizations use "Research" as the place to evaluate newly hired Ph.D.s with the understanding that only a small fraction will stay in Research. This is a way of easing the transition from the university and facilitates effective technology transfer.

Reorganization, mergers, or consolidation of organizational units pose a particular set of problems. The best guideline is to start with an assessment of the future objectives that have prompted the change, without reference to existing capabilities. Next, define the optimum capabilities and organization required to address these objectives. At that point, the attempt to match existing resources to future requirements will serve to define a set of constraints, principally the time required to implement desired changes.

Proponents of solutions based upon a priori (or worse, ab initio) concepts, such as central versus distributed functions and project versus matrix organization, should be viewed with suspicion. It is my belief that the creative effort associated with exemplary R&D flourishes in the most adaptable, least rigid organizations.

Returning to the question of resource allocation, it has been my experience that the question: "What would you propose to do with a 20% increase in resources?" most frequently is answered by: "More of the same." This is not surprising. It is difficult to have a commitment to what you are doing without believing in its importance. What is more, to suggest something else may well threaten existing fiefdoms. This is the basis for the earlier statement suggesting the need for proposals addressing "new" prospects.

The question: "What activity would you drop in the case of a 20% cut?" is likely to be more useful in considering the amount of R&D appropriate in a given situation.

Finally, the responsibility for the decision to increase or decrease R&D is necessarily with top management. Setting of priorities within the budget should be the responsibility of R&D management.

For example, an effective way of dealing with the question of basic versus applied research in industry has it that the Director of Research never requests resources for basic research, explicitly. It is understood that a certain percentage of any resources for new, applied projects are allocated to basic research. This helps maintain the balance which has been found effective. It also avoids the difficulty of having management outside of Research make decisions which are likely to be beyond their competence.

Acceptance of such responsibility on the part of R&D management is consistent with the desire to maintain an atmosphere conducive to creative effort. Reluctance to make hard choices and accept the associated risk is a sign of weak management. This is particularly true for the R&D function. The alternative is to have someone less qualified set priorities for you.

9.8.4 Setting Priorities

(This is the hard part!)

Setting priorities is making decisions about what will NOT be done. The project which has been approved will itself produce additional insight. This insight will be the basis for corrective action. Not so with the project which has been shelved. No one is held accountable for the shelved nonproject.

Useful generalization:

Before a decision is made, the lower priority alternative deserves the more careful attention. There is a tendency in collective decision making to identify with the winning side.

Many of the problems associated with setting priorities are not of a scientific or technical nature. The national priorities of building a super-colliding accelerator, of proceeding with a human genome project, or building a space station are not likely to be decided on their scientific merit. It is essential that the R&D manager understands why his or her overwhelming arguments have failed to persuade the decision makers. The problem is not the lack of relevant knowledge on the part of those listening. It is the presenter's failure to appreciate the basis of the listener's concerns.

The decision to emphasize R&D on substances which are not carcinogenic may be a better long-term bet than to work on remedies for the effects of carcinogenic materials. Yet, given a breakthrough in a promising cure for cancer, one would not hesitate to shift applicable resources to such an effort in order to effect timely transfer. The crucial issue may well be the definition of "applicable resources."

Some high-level policy makers are known to have the notion that one can "reorient" R&D efforts on the basis of perceived similarities in skills. A group of physicists, who have spent the past 10 years in the field of nuclear energy, are unlikely to produce much of value in the field of semiconductor physics. Both groups may refer to themselves as physicists, but their qualifications are as different as those of a cardiologist and an orthopedic surgeon. Surely, both can be retrained over a period of time, but at that point, they are beginners in the field, hardly in a position to compete with those who have many years of task-relevant experience.

Useful generalization:

In the long run, it is easier to shift priorities in an R&D function which has a diverse mix of disciplines. The nucleus of those appropriately skilled can help those to be retrained become effective.

One of the most insidious problems that interfere with setting meaningful priorities for the R&D function is the so-called "Not Invented Here" or NIH syndrome. This syndrome can be born of false pride, where we cannot bring ourselves to admit that we cannot be first in everything. Worse, it can be based on sheer arrogance, i.e., "If it were any good, we would have invented it." On the surface, this problem would seem to be confined to the commercial world of industry. But there is plenty of evidence that it also exists in university and government laboratories as well as in pertinent government agencies.

More often than not, the problem emanates from the top. As we suggest in a subsequent discussion of cultural values (Sect. 13.5), it is top management that sets the values for the organization. To read in the house organ of a leading electronics firm that a major competitor's product line is the fastest growing market segment in the industry gives a clear message: We have now picked up this idea and have a product which improves upon the original!—2 years earlier, the NIH syndrome had served to deny this project the priority needed to bring it to market.

Useful generalization:

NIH is counterproductive. Most creative work builds on efforts and insights that have gone before.

9.9 Conclusions

To be effective, the process of evaluating must be continuous, be it personal performance, projects, or an entire R&D function. The process must be structured in such a way that the people affected by the results of the evaluation perceive it as valuable to themselves. To be perceived as valuable, an evaluation must lead to decisions, followed by action. It can be a most valuable management tool.

For me, participation in project and functional evaluations has been the single most effective management training. Task forces established for the purpose of evaluation should include the widest possible range of experience. The least satisfactory evaluation team is comprised of a group selected for their common interests and recognized "dependability." Nor does such a group provide much of a learning experience for its members. The benefit of diversity among task force members is well illustrated in Nobel Laureate Richard Feynman's account of his participation in the investigation of the disaster which befell the U.S. space shuttle "Challenger" (Feynman 1988).

9.10 Summary

The evaluation of R&D efforts has been considered first from an operational point of view, reflecting the responsibilities of the various levels of management.

The project manager must be particularly concerned with problem identification and outside dependencies. The next level of management must consider the consequences of increased or decreased project resources in a larger context. At the highest level, there is the added consideration of possible impact upon the strategic direction of the organization. The basis for a given strategy, be it market-, technology- or finance-driven, is a factor in the evaluation at the top level.

Attention is called to the importance of "Bootleg" or "Under-the-Table" projects and the associated need for slack time.

The manner in which evaluations are conducted can have unwanted effects. This is analogous to the problem in physics where the method of measurement can affect the quantity to be measured. Some criteria for evaluating R&D efforts are suggested.

Periodic evaluation of ongoing projects increases the evaluators' understanding of the critical factors affecting the project. Evaluation of new projects or proposals is more difficult since the basis for judging the conditions for potential success is more limited. Also, the imagined relevance of past experience may distort the evaluation.

Evaluation of an entire R&D function takes a more global view. It focuses on the impact of R&D upon the sponsoring entity. The complexity of the issues and the required breadth of knowledge and experience serve to point up some of the difficulties one encounters. The "Useful Generalizations," which have appeared in the above section, are brought together here for ready reference:

Useful generalizations about the evaluation of R&D functions:

- There is little to be gained from an evaluation of the success potential of an activity that is judged irrelevant to current or future organizational objectives.
- An evaluation must be based upon a set of criteria clearly understood and agreed upon by the evaluator and the evaluated. Such understanding and agreement must be reached at the time an activity is initiated, not just before or during an evaluation.
- The evaluator must have a clear understanding of the motivation for the evaluation on the part of those initiating it.
- It must be understood by R&D management that the burden of making their function relevant to the entity served is clearly on their shoulders. This means developing a meaningful way to evaluate the function *and* responsibility for effective technology transfer.
- Before a decision is made, the lower priority alternative deserves the more careful attention. There is a tendency in collective decision making to identify with the winning side.
- In the long run, it is easier to shift priorities in an R&D function which has a diverse mix of disciplines. The nucleus of those appropriately skilled can help those to be retrained become effective.
- NIH is counterproductive. Most creative work builds on efforts and insights that have gone before.

10. Administrative Skills

Making presentations and conducting meetings are important skills of a manager. These skills are equally valuable to the nonmanager in the Hi-Tech environment, where a lot of everyday business is conducted in meetings and through presentations.

10.1 The Art of Making Presentations

Formal presentations have entered our culture to the point where children are encouraged to make verbal presentations in grade school, and plays such as "How to Succeed in Business Without Really Trying" spoof the overly aggressive sales pitch.

After both giving and listening to presentations for many years, one develops strong reactions to presentations which fall short of the expected. Here are some of the elements of a good presentation, relevant to the Hi-Tech environment, i.e., not necessarily relevant to door-to-door selling.

10.1.1 Know Your Audience

The most common failure of presentations aimed at audiences in the Hi-Tech environment is an inadequate appreciation of the background and expectations of the people who have come to listen. To illustrate:

Bob asks me the day after his presentation, "How did it go?"—My response: "Terrible!"—He protests: "But I have had nothing but praise when I gave this same presentation the previous time."

Upon inquiry, Bob admitted that he did not know anything about the group of people whom he had addressed on this occasion.

Once again we must consider our objective, the results we want to achieve, in making a presentation, i.e., not any presentation, but this specific presentation to this

R. Kay, *Managing Creativity in Science and Hi-Tech*,
DOI 10.1007/978-3-642-24635-7_10, © Springer-Verlag Berlin Heidelberg 2012

specific audience. Presumably, we want to convey the information most relevant to the objective of our presentation: Meet or exceed the expectations of the audience.

To spend a lot of time on a tutorial introduction to the subject when talking to experts is an affront and will be perceived as unwillingness or inability to deal with the tough or novel aspects of the subject which the experts have come to hear.

In contrast, a presentation on the same subject to a group of lay people should place the major emphasis on the tutorial introduction. This will prepare them so that they will understand this to be an area of potential interest (they will decide early on in the presentation). They may thus anticipate that something is to be reported that will increase their interest in this subject.

Of equal importance to the background of the listeners is an understanding of their expectations.

- Why are they listening to this presentation? To evaluate the presenter? To evaluate the project being discussed? Evaluation relative to what?
- Has the audience spent the previous 5 h listening to similar presentations? To a presentation on the very same subject just before? What have they heard already?

When asked to give a presentation, one should always demand to be provided with the background and expectations of the audience, as well as with the complete agenda if the presentation is one of several this audience will listen to on this occasion. If such information is not available in advance, it is entirely meaningful to ask the audience:

- What brought them to your organization?
- What is the purpose of their visit?
- Is there anything specific that they want to learn or find out on this occasion?

Failure to do this puts the presenter at risk of wasting the audience's and her/his time and of leaving an unfavorable impression. Assuming that the presenter is familiar with the subject, it should be possible to tailor the presentation and shift emphasis for maximum relevance to the audience. There is nothing more frustrating to both the presenter and the audience than a mismatch between the content of the presentation and the interests of the audience.

10.1.2 Presentations: Structure and Technique

Any presentation of more than 5 min should start with an outline of what is to be covered and the presenter's purpose.

The following examples are the introductions to three presentations made about the same project. Each addresses a *different* audience and has a unique purpose.

Example 1. This would be an appropriate introduction for a presentation to a group of people working in the general field, but with only limited familiarity with the specific topic. They are assembled to consider the presenter for a position.

This is a presentation of some aspects of my thesis. My thesis is concerned with a new methodology for providing scientists and engineers with a programming environment for applications uniquely able to exploit a new class of parallel computers. The applications which are the focus of my work involve the solution of differential equations by the multigrid method. After briefly defining the hardware and the range of applications which can make use of the multigrid method, I will concentrate upon the aspects of my work which are thought to be new and unique in facilitating such application development. I will also discuss some of the alternatives which have been evaluated and discarded. I will attempt to convey to you the manner in which I have approached this project and the assistance I have received from Prof. Hotshot and my colleagues.

Example 2. This is an introduction to the same topic addressing the monitors of the grant which supports the project. It is the first of several presentations on the Parallel Computing Project.

I will report the current status of the aspect of the Parallel Computing Project concerned with creating an appropriate programming environment for engineering and scientific application development. There will be a brief discussion of current thinking on the range of hardware and applications to which this environment is applicable. There will be an attempt to show the relationship of the Programming Environment to the overall motivation and aims of the Parallel Computing Project. I will start by referring to the project organization and the current status of our schedule and budget. I will conclude by indicating the specific accomplishments since the last review and the projected activities for the next year.

Example 3. A presentation to a group of congressmen, who are reviewing projects aimed at maintaining leadership in advanced computing, may go like this (it is assumed that this will be the only talk on the Parallel Computing Project they are going to hear):

I will try to make clear the overall motivation for this project by showing how Parallel Computing is the most promising approach to maintaining leadership particularly in scientific and technical applications of computing. There will be a discussion of the role of and interrelation between the development of special hardware and software for parallel computing. I will attempt to show the unique problems of programming in this new environment compared to the more conventional programming which has been around for many years. I will conclude by making reference to related efforts in this area and try to assess the potential impact of our work.

In each of the above instances, the focus is upon the interest of the audience and takes into account their background. The presenter has a clear purpose in making the presentation. Eighty percent of each audience should feel that they have understood 80% of what they have heard. No one in the audience should feel that he/she was left behind after the introduction or that he/she was talked down to.

Attempts at humor or flippancy tend to detract from a presentation which is to be taken seriously. It is a great art to bring humor to a serious topic; your professional standing is likely to be jeopardized by appearing as an amateur comedian.

The presenter's mannerism and appearance should not distract from the presentation. This means it should reflect the norm for the given audience and environment. This brings to mind the following incident:

One of the brightest and most creative people in the laboratory asked my advice about a pending presentation. Both the subject and the audience were very important to this young man. I had a great deal of respect for him, in spite of the fact that he affected unconventional manners, something one is more likely to accept in exceptional people.

At some point in our discussion, he asked if I saw something wrong in his wearing some outlandish jacket for this upcoming presentation. He was surprised by my lack of reaction to this intended provocation.

Upon his prodding, I tried to make clear to him that in wearing that particular jacket he would, no doubt, succeed in distracting his audience from the subject of his presentation. If that was his intention, I thought the jacket was very appropriate. (On this occasion, I made a life-long friend. One of the rewards of management.)

Only experienced speakers can effectively deal with questions during the course of a presentation without being thrown off the track or time schedule. To run out of time before the all important conclusion is equally frustrating to the audience and the speaker.

10.1.3   Visual Aids

Visual aids generally are very useful in guiding all but both the most experienced presenter and the audience. The aids should be appropriate to the occasion, not too glossy but not sloppy either. Above all, they must be readable. There is no excuse for excessive material on a given page, lack of attention to good formatting, or dirty or much-handled material. Remember, the audience has nothing else to look at for an hour. This is a display of your attitude toward the subject and toward the audience.

The audience may be sympathetic to an inexperienced presenter under pressure of an audience of unfamiliar faces. But poorly prepared visual aids are perceived as lack of interest or regard for the audience.

Power Point presentations, which have become the norm in the Science and Hi-Tech environment, have largely overcome the problem of formatting for maximum legibility.

Be cautious with the use of special effects such as animation. These can be very effective if properly used. But you do not want to obscure your message by having the audience focus on your dazzling visuals.

The time spent by the audience listening to your presentation represents a significant investment on their part. Lack of preparation on the part of the presenter is inexcusable.

Presentations in a totally darkened room should be avoided. A more than minimally dimmed room not only promotes nodding on the part of inattentive listeners, but also deprives the speaker of the important eye contact with the audience. Without eye contact, it is easy for a speaker to lose an audience.

10.1.4   The Art of (Not) Listening

Many of us find ourselves sitting through an occasionally boring presentation, or a presentation which in one way or another does not meet our expectations. Rather than show boredom or displeasure, a more positive stance would be to use the opportunity to analyze the strength and weaknesses of the presentation itself—a most effective training vehicle.

Administrative skills such as making presentations will contribute significantly to your effectiveness. Like good study habits, the sooner they are acquired, the more benefit can be expected. There are numerous books which go into every aspect of presentations, e.g., Ascheron, Claus and Kickhuth, Angela (2004).

After consciously listening to a few presentations with the aim of learning what is effective and what is not, you will have a basis for testing how these ideas work for you.

10.1.5   Summary

Factors which influence the success of a presentation are as follows:

- Appreciation of the audience's expectations
- Match between speaker's intent and audience expectation
- Speaker's awareness of the relevant background of the audience
- Introduction and outline to convey to the audience the speaker's objective in making the presentation
- Avoidance of detours into presenter's favorite details or virtuoso passages
- Attempts at humor by people who are not funny
- Summary and conclusion to emphasize the major points the speaker wishes to leave with the audience
- Appropriateness of visual aids

10.2   The Art of Conducting Meetings

The need for and value of meetings in the discharge of managerial responsibility have already come up in connection with running a project, appraising performance, and as an effective form of communication in general. It is the objective of this section, not only to present some proven techniques of conducting meetings, but also to raise awareness of some of the factors that can influence their effectiveness.

10.2.1   General Guidelines

Someone must be responsible for calling and conducting a meeting. If an event, such as an emergency, triggers a meeting, someone must take responsibility for running the meeting.

This may be the senior person, senior in rank, age, or experience. Alternatively, it may be the person who has most at stake from the outcome of the meeting.

It is up to this person to make everyone aware of the purpose of the meeting, if not prior to the meeting, at least when everyone is assembled. It is unreasonable to expect someone to attend a meeting without understanding its purpose.

A general acceptance of the need to attend meetings as a matter of routine should be challenged from time to time. I have seen numerous occasions when people have attended meetings for a variety of reasons, not all of them valid.

If the meeting is informational, one's need to know should be questioned. If the purpose is to make decisions, there is no need for spectators. People often rationalize their attendance under the pretext "... if something important should come up, ..." when in fact they feel insecure about their position or role in the organization.

Thus, the person calling a meeting must take the responsibility for committing a valuable resource—other people's time.

The ability of the attendees to contribute effectively will, to a large extent, depend upon the preparations they have been able to make.

Since people attending meetings presumably have other responsibilities, the meeting should be scheduled a reasonable time in advance. Notification of the expected length of the meeting is important for the same reason.

10.2.2   Sample Meeting Agenda

Potential problems of attendance and scheduling can be avoided by the use of a standard form for announcing meetings, such as the following:

To:	List of names of people asked to attend
From:	Person calling meeting
Subject:	Design of new Chemistry Building
Date of Meeting:	March 17, 2020
Time:	2:30–3:15 PM
Place of Meeting:	Campus Math Building, 10th floor, conference room B.
Purpose:	To decide location and overall size of building

The inexperienced may perceive this as an unduly bureaucratic procedure. But if a significant portion of time must be spent in meetings, not an unusual situation for managers in Hi-Tech enterprises, a disciplined approach will be appreciated by all involved. Without it, one frequently finds oneself at odds with colleagues over scheduling and procedural issues, which could be avoided.

If for some reason it is necessary to leave a meeting before the scheduled adjournment, it is appropriate to inform the person presiding beforehand. The implication of early departure is that you are a very busy person with more important things to do, but that the others are not. It is up to the person running the meeting to excuse you to the other participants if called for.

If progress or decisions are held up through the unscheduled absence or late arrival of one of the invited participants, those attending will feel that they are forced to waste time.

The person calling the meeting must decide who should attend. The objective of a meeting is rarely advanced by having other than potential contributors in attendance. When decisions are to be made, or are expected to follow from the meeting, those responsible for the decision must be involved. The person calling the meeting may indicate when a representative of the invited person is or is not acceptable.

Instructive is the procedure used by a major corporation in determining attendance at decision-oriented meetings of its highest level of management.

The handful of people who are ultimately responsible for the decisions clearly do not have the time or knowledge to become familiar with every aspect of the wide range of issues they must consider and approve. The executive among the decision-making group who has functional responsibility for the issue up for decision is already familiar with the content of the presentation and has approved it.

An extensive staff operation, headed by senior managers with operational experience, must review each recommendation that comes before the top executives. Nonconcurrence at lower levels or between the recommending party and the staff is expected to be resolved before the meeting. The only staff or affected parties asked to attend the meeting are those who have an objection which they have not been able to resolve before the meeting. Once in attendance, they must state their objection and the reason for lack of prior resolution.

Thus, in these meetings, top management's concerns with policy and strategy are brought to bear upon decisions which have been approved at the operational level or, in case of unresolved conflict, are addressed by staff attending for that purpose only.[1]

10.2.3   Methods for Conducting Group Meetings

The most common and useful meetings are periodic staff meetings where project personnel discuss current problems. The manager responsible for the conduct of

---

[1]It is appropriate to mention that this format was abandoned after serving well for more than 10 years. Line management had learned to game the procedure by manipulating the system based on line management/staff contention.

the meeting can exert considerable influence, positive or negative, by choosing the format of the meeting.

For example, as newly appointed manager of a design group which has encountered a major obstacle, you are faced with several issues: You do not know the capabilities of the group, what has been tried before, or the basis for the problem encountered. You want to address these issues in a meeting including all of the project staff. There are various ways of approaching this:

Brainstorming: In this mode, everyone is encouraged to come forward with ideas, which are recorded as presented for later evaluation. The objective is to get as many raw ideas in front of the group as possible. An idea put forward often stimulates another.

This approach can work well when the participating members are in the same peer group, i.e., there is no feeling of senior/junior or professional/nonprofessional distinction. Otherwise, one group will feel inhibited or at a disadvantage. It also places the shy, less secure, and more reticent members of the group at a disadvantage.

Brain-writing: This is a variation of brainstorming in which each person around the table puts his/her idea on a piece of paper, and passes the paper on to the next person to enter the next idea until people run out of ideas.

This method preserves the advantage of being stimulated by the ideas of others, and overcomes the disadvantage of intimidating some of the members of the group. It lacks some of the spontaneity of verbal brainstorming, which is often the driving force in such sessions.

Structured Brainstorming: The discussion leader solicits ideas from each participant in turn, going around the table. Except for its lack of spontaneity, it has all the advantages of brainstorming and assures a hearing for everyone. It is not a bad idea to start out with the person next to the most reticent member of the group who would thus get the second turn. The success of this approach depends upon the leader's ability to solicit immediate discussion of each idea without exhausting the subject. Otherwise, the leader will preempt contributions from others.

Evaluation of Prepared Positions I: Before the meeting, participants have had the time to prepare a response to stated questions or problems. They may all be asked to respond to the same question, or to different questions relating to a given issue.

For example, what will be the effect of a 50% cut in network access? State potential impact upon schedule and what function or component you would drop in order to still maintain the current schedule.

This approach is likely to produce better results than the unprepared response to such a question raised in a meeting.

Evaluation of Prepared Positions II: In preparation for a meeting, participants are asked to make recommendations on how to deal with an issue which has broad implications. Here, the problem has been defined broadly, not in the form of specific questions.

For example, what is the single most effective way for this project to speed up the schedule by a factor of two?

In this case, the manager is trying to elicit a range of responses, some conservative, some unrealistic, some imaginative, and some pedestrian.

Evaluation of Prepared Positions III: The manager comes to the meeting with all the possible responses to a question that have occurred to him/her, or that he/she has been able to consider. The participants are then asked to vote on their preference. This approach may be useful if the manager wants to make a decision within a set of constraints and take the desires of the majority into account.

For example, there are four time slots during which the group must take its vacation during the coming year. We must decide upon a given time slot for the entire group. In this case, the manager is responding to a real constraint imposed from the outside.

Or, here are the criteria I will use in selecting a representative of the group to attend the International Software Engineering Conference in Monte Carlo. Would you express your preference by ranking these criteria?

In this case, the manager is not willing to consider criteria other than those proposed. Otherwise, the manager could have elicited criteria from the group by brainstorming and subsequently ask for the ranking.

10.2.4   Selection of Method

There are obvious variations on this theme. Two principal considerations enter the selection of the most effective method:

1. Which approach appears most appropriate to the situation and group of people involved?—The *task-relevant maturity*[2] of the group and the uniformity of its distribution among the participants are one of the considerations. Is the group to be involved in the decision, or is it merely to recommend a course of action?
2. Which method is the manager most comfortable with? Is the resolution of the issue likely to provoke reaction that will have broader implications? Does the manager need approval of the decision at a higher level?

Experience is probably more effective than analysis in developing confidence in one's judgment in this matter. This confidence is attained more readily by those who are able to benefit from the experience of others. Every time you encounter such a situation as a participant, try to judge its effectiveness and the reaction of the people involved.

---

[2]The notion of "task-relevant-maturity" was brought home to me on the occasion of a management workshop organized by a European National Physical Society attended by Ph.D. candidates and young Ph.D. physicists from industrial and government research labs. They were divided into groups of 10, given a problem to solve and told to select a spokesperson to present and defend their consensus solution. Less than half of the groups were able to complete the task. The reason: They were unable to agree upon a spokesperson!

10.2.5   Tips for One-on-One Meetings

A manager should routinely have a weekly, one hour, one-on-one meeting with each of his/her subordinates. The subordinates should feel that this is their meeting. The manager must make every effort to help the subordinate take initiative in such a meeting.

Preferably, such meetings should be held at the workplace of the subordinate. The manager can learn a lot from the environment which the subordinate has created or is subjected to. This also avoids interruptions due to other demands made upon the manager. It increases the manager's visibility among the people he/she is responsible for and helps create an image of being there to help the people reporting to the manager. While some managers argue the efficiency of having people come to their office, this may be achieved at a sacrifice in effectiveness.

If the subordinate has some personal concern, it may be postponed until the end of the meeting. The manager must be sensitive to this very human trait and not cut the person short just as they get the courage to raise a sensitive issue. I have had the repeated experience of a person having come to my office and hesitate upon leaving, with the door knob in hand. Unless asked if there was something else they wished to discuss, they would leave. More often than not, they will bring up a personal issue, which they find difficult to discuss.

If, as manager, you want to be the first to know about potential problems, a closed office door, and/or one guarded by an overzealous secretary does little to enhance your availability. E-mail has done a lot to alleviate this problem. If you frequently ignore messages addressed to you, one of the potential advantages of the system, this source of input will dry up.

While a written agenda is not appropriate for a one-on-one meeting, the manager may invite the subordinate to start out by enumerating the items he/she wants to cover in this meeting. If this is done regularly, subordinates recognize that they are responsible for the agenda. If the subordinate does not have any specific items for discussion, it is still important that he/she be encouraged by way of questions. It is all too easy for a manager to take the initiative and do all the talking.—Beware!

One of the satisfactions of management comes from the feeling that you have been effective in helping someone: helping someone to overcome or avoid a problem; helping someone to do a better job; or helping someone to better realize his/her personal objectives. The one-on-one meeting is the starting point.

Some managers keep a daily journal and choose to keep a record of such discussions, particularly when they are trying to resolve a problem. Regular one-on-one meetings through the course of the year are the best assurance of an effective performance appraisal. What is more, it is desirable to solve small problems as they arise, rather than letting them become large problems through neglect. This may sound trite, but avoidance of conflict is an all too human trait.

10.2.6 Summary

In calling for a meeting, preparing for a meeting, or participating, the person responsible must recognize and convey to everyone involved that valuable time is being taken from other activities.

Careful preparation and consideration of the most effective approach toward a particular problem can go a long way toward a successful meeting, one where everyone leaves with a feeling of accomplishment. The need to be sensitive to problems which are difficult to broach in one-on-one meetings is pointed out.

11. Starting a New Enterprise

Creativity in the Hi-Tech world has found no better expression than in the start-up of a new venture. Entrepreneurship and innovation have been the basis for much of the creative activity we are trying to manage. A look at what it takes to start a Hi-Tech enterprise provides insight into some of the unique aspects of managing creativity.

11.1 Profile of the Entrepreneur

The entrepreneur who sets out on a new venture is the key ingredient in its potential success. If the initial concept is wanting, it is possible to revise the business plan or change the product mix. Only the individuals themselves who have staked their future on the new venture are given.

What then, are the desired qualities of an entrepreneur? This subject has been pursued in great depth by serious scholars such as Drucker (1985b) and by groups such as the MIT Sloan School of Business. The subject also receives coverage in journals which address themselves specifically to the recent or would-be entrepreneur. What can we distill out of this material which is uniquely applicable to the Hi-Tech entrepreneur-to-be?

1. Many able entrepreneurs have gained valuable experience in the corporate world. Drucker (1985b) goes so far as to state categorically that some corporate experience in management is one of the essential qualifications of the successful entrepreneur (dramatic exceptions to the rule, notwithstanding. RK).

Yet, venture capital firms tend to be apprehensive about the would-be entrepreneur who has been successful in a large corporation. This caution is based on the concern that such people are used to extensive staff organizations, which provide much of the basis for decision making, and that they are therefore ill prepared to meet the demands of a start-up operation.

This seeming contradiction is resolved by recognizing that the concern of the VC firm applies principally to people who have advanced to senior executive positions.

2. "The true entrepreneur has a decisive nature and loves to act on instinct. These rather bold types are generally loners who can relate well to people when they so choose" (Brown 1980).

While great achievements in the Hi-Tech world are generally associated with superior analytical ability, it is rare to see such ability combined with a great trust in instinct. Successful Hi-Tech entrepreneurs are more likely to pursue an idea based upon what may be inadequate knowledge and experience, see how far they get, and then seek help or alternative approaches to solve the problem.

The willingness to make conclusive decisions, i.e., to take action in the face of uncertainty, is the principal mode of entrepreneurial operation. Indecisiveness, often disguised as the desire "to get more adequate information," is incompatible with an entrepreneurial personality.

The successful entrepreneur invariably has the leadership qualities which are required to inspire commitment and loyalty in those who join up. Enthusiasm and endurance best describe these personality traits in the entrepreneur.

3. Money, position, and status count principally as the measures of entrepreneurial success. They are not ends in themselves. Creating something and getting something accomplished must dominate the start-up phase of a new enterprise. When selecting partners or key associates for a start-up venture, those are the ambitions to look for.

4. A consummate salesperson or marketing genius, often disguised as engineer, scientist, or financial expert, is the type of person frequently found among successful entrepreneurs. Since what you are designing, building, or selling is generally new or addresses a newly recognized need, your salesmanship is an essential part of communicating with potential customers and suppliers of the goods, services, and the investors upon whom the new enterprise depends.

The personality traits which go along with successful entrepreneurship also have their negative side. In a study of 38 entrepreneurs Kets de Vries (1985) observes: "The boundaries between very creative and aberrant behavior can be blurry... The mix of creative and irrational is what makes many entrepreneurs tick and accounts for their many positive contributions."

The consequences of such behavior are most relevant to those who are dependent upon frequent personal interaction with the entrepreneur. A word about partners in the founding of a new venture. It would be a good idea to have potential partners write down their personal objectives in starting the enterprise. This not only makes interesting reading, but also is an excellent way to determine potential conflict among individual objectives. If more than one of the partners places prime emphasis on being the boss, there is trouble ahead. If the various members all want to be responsible for defining the "architecture" of the system, more problems.

True entrepreneurs are more likely to start on their own and later hire people who complement their talents.[1]

The above generalizations apply to many successful entrepreneurs. But only exceptional people will produce exceptional results. What makes them exceptional in producing consistently worthy results (and worthy to whom) can only be judged a posteriori. Bill Gates or Mike Zuckerberg is not likely to take the time to read a book about management.

11.2 The New Enterprise

It is not unusual for two fellow MBA students, after some few years of diverse experience, to be drawn together by a mutual desire to start a business of their own. This is to say, it is not a specific business opportunity or product idea which is the basis for the start-up. They are about to explore available opportunities which offer promise, based upon their expertise or experience.

This scenario is not typical for Hi-Tech entrepreneurs. They are more likely to come from a medium or large organization, which is unable to pursue an idea which looks promising to the would-be entrepreneur(s). This may be due to the fact that resources for new endeavors are always finite, that the idea does not fit the manufacturing or marketing resources of the organization, or that the projected market is too small relative to other product opportunities, etc.

"In a large company, every product must be a home run to justify the cost of marketing and development," Jack Goldman formerly *Xerox* research chief's way of saying that there is a certain minimum cost associated with development and marketing in a large company, regardless of the revenue potential of the product (Uttal 1983).

These are usually good reasons for the particular organization to pass up a given opportunity, without impugning the idea itself.

Other opportunities arise from the conviction of the would-be entrepreneur that a potentially better approach to a given product or service is available, but is competitive with the approach of the existing organization.

While the exceptions get more than their proper share of attention, most successful Hi-Tech start-ups are aimed at the fulfillment of an unsatisfied need or a novel way of satisfying a need. That is, they are market rather than technology driven. There is nearly complete agreement that the recognition of unfulfilled customer needs rather than new functions made possible by new technology is the basis of most successful new ventures. (Note: we are talking about new ventures, not new products or services.)

It is true that the technological feasibility of dry copying created the office copying market. Earlier technological approaches did aim at low cost and convenience.

---

[1]It should be instructive to review the "Profile of the Creative Individual" in Sect. 3.3 to recognize the potential dichotomy between the "Entrepreneur" and the "Creative Person".

But they never satisfied these needs to the point of opening up the market. Leasing the relatively expensive first machines and the "per copy charge" proved to be the key to success.

The development of spreadsheet programs did as much for the popularity of personal computers as the other way around. One factor in the early popularity of these programs is that they allow people with limited typing skills to do an effective job of data entry. It satisfied an unfulfilled need of every professional with limited typing skills.

> This suggests that it is important for the new venture to have among its founders someone who understands the needs of the targeted market and has access to the potential customer set.

11.3 The Business Plan

Earlier on, we emphasized the importance of a written strategy and operating plan. It provides a structured and thought-through set of objectives and a plan for their implementation. In starting a new venture, the business plan serves this same purpose. It helps to establish consensus among the founders of the start-up operation. A business plan is a prerequisite to raising start-up funds.

To do without a formal business plan is possible only to the extent that start-up capital is available from sources willing to accept the would-be entrepreneurs on faith. Such sources are the exception and are usually limited to family and friends. Venture capital sources and banks, or in other words, people who assume responsibility for investing other people's money, require a business plan in order to verify claims and assumptions.

In established enterprises, the evaluation of new development projects or product ideas is based upon a proposal which involves the same considerations as the business plan in the start-up venture.

The following pages provide an example of the executive summary of a business plan. The particular emphasis of this summary is an important aspect of its effectiveness. It is designed to provide the investors with a basis for judging their potential interest. For a detailed discussion, see Zaytsev (2005).

11.3.1 Executive Summary of Business Plan

Company: MegaComp Systems

2001 Century Blvd.

Los Debug, CA 95030

(418) 444 6664

Contact: Steve Sharp, President

Type of Business: Software products and services for users of computing-intensive applications. Emphasis is system software for multiprocessor and parallel processing hardware configurations such as those manufactured by the Galaxy, Universal, and V-Cube Corporations.

Company History: The company was formed in 2010 to develop, market, and service software that addresses both general and special needs of computing-intensive applications.

Management: The two principals were formerly with the ABC Parallel Systems Project, one of the leading university groups in the field.

Dr. Steve Sharp, President, has been in the software development field, both in the university and in industry for the past 10 years. He was responsible for guiding the ABC Unix project, which has become the operating system of choice among sophisticated end users. He was instrumental in transferring the results of the work done at the University to the Univocal Unix Corporation, (UUC), which currently does a $40 million business a year and has captured 30% of the worldwide market for systems serving this growing industry. He is on the board of directors of UUC.

Laurel Aldner, Senior Vice President, was Vice President, Marketing, of UUC. She was responsible for the development and implementation of the marketing strategy which led to the successful distribution of the ABC Unix system through UUC. In her prior role with the ABC Unix project, she was responsible for the design of the screen and menu manager, which has been the central feature accounting for the popularity of the system with sophisticated end users.

Senior sales and financial personnel will be added as soon as possible to complete the management team.

Product: The operating system and inter-module communication handler for the V-Cube family of computers is designed to provide an open-ended application development environment. MegaComp Systems has an agreement with the V-Cube Corporation which provides for an exchange of preannouncement specifications and early access to hardware and software, respectively, to effect an optimum test environment and access to early users of the system.

The initial release of the system will include an editor, file management, and interactive debugger aimed at computing-intensive applications.

Compatibility with Galaxy and Universal hardware systems is planned for early release.

Future releases will include a communication interface to IBM and HP main-frame installations and a terminal server designed to take advantage of the new cloud based capabilities.

The planned software offerings are designed to establish MegaComp Systems as a major supplier of software to the large scientific and engineering user community. As the parallel computing hardware achieves greater penetration and the lower prices associated with larger volume, MegaComp expects to share in this larger market.

Market and Competition: There are presently about 150 supercomputer centers operating in the Western world, each supporting about 1,000 technical users. The centers are run by government (50%), industry (40%), and universities (10%). This represents the initial customer set. The introduction of the V-Cube hardware will reduce the entry cost of a supercomputing facility by a factor of 50 over currently available hardware systems.

This is expected to provide a potential market for MegaComp software of 7,500 installations over the next 5 years. We expect to command a 40–60% market share or 3,000–4,500 installations in the supercomputing facility class.

By the end of our third year of operation, the hardware cost is expected to drop by a factor of 3, thus raising the potential number of installations to about 10,000. While this is a large number of installations, it is but a small fraction of the general purpose computer installations, which are estimated to reach about 200,000 in the next 10 years.

This reduced cost of hardware will bring the cost of such systems within reach of all large and many medium-size engineering, architectural, and construction firms. We also expect that the introduction of Data Mining and Knowledge-Based systems in financial institutions will draw upon the capabilities of parallel computing hardware. We consider this a very promising market, but have not included it in our financial projections for the first 3 years. We have planned a small effort to explore opportunities in this area in our second year of operation.

Our agreement with the V-Cube Corporation provides for some joint marketing effort. We expect to have our own sales personnel make joint calls with V-Cube, in order to develop an in-depth understanding of future customer requirements. We shall also pursue potential customers who are planning on other than V-Cube hardware.

Based upon our assessment of the superior cost/performance of the V-Cube hardware, our agreement with V-Cube will ensure us a unique competitive position. We expect to enhance this position by our ability to attract some uniquely qualified people, by virtue of the technical challenge we can offer to outstanding personnel.

The size of the initial market and the level of sophistication required of early customers will not make this an attractive opportunity for the well-established, large software firms for some years to come. We expect competition to arise from other start-up firms who have chosen to aim their software primarily at other than V-Cube hardware, but who will eventually seek general applicability. None of the start-ups currently known to us appear to have the level of relevant experience we expect to bring to bear.

Financial Projections: The 5-year plan calls for the following annual revenues and after tax profits. All figures are in thousands of dollars.

Year	2012	2013	2014	2015	2016
Revenues		1,250	5,000	30,000	60,000
Net income	(300)	(200)	1,000	6,000	15,000

The business plan itself (not included here) must provide extensive backup and answer as many questions as possible without going into unnecessary detail.

It is important that the person verbally presenting the plan to potential investors be familiar with the basis for every explicit statement in the written plan. The most frequent shortcoming of Hi-Tech business plans is an excess of technical detail about the product and a shortage of relevant market and financial information. The investor's initial interest will be based upon the anticipated growth potential. Only such potential will prompt detailed analysis of the technical merits of the plan by experts available to the investors.

Likewise, information about the principals should focus upon that part of their experience which is relevant to the new venture. The principals are the most important assets as far as the investors are concerned. A detailed account of the experience and educational background of the principals is usually attached to the summary.

Since a significant portion of investment capital for start-ups comes from the venture capital industry, it is incumbent upon the would-be entrepreneur to become familiar with the operation of that industry. (An overview of the venture capital industry, including a section on the evaluation of business plans, can be found in Chap. 12.)

11.4 Financial Controls

An adequate understanding of financial controls requires familiarity with the subject of finance and accounting. Here, we do not assume this understanding, and hence must focus on a few issues crucial to the subject. We are also trying to motivate current and potential Hi-Tech managers (and would-be entrepreneurs) to give this subject the attention which it requires. A typical MBA program would cover accounting and finance in separate multiple semester courses, based upon a preliminary exposure to economics and statistics.

> By far the largest number of failures among Hi-Tech startups is the direct result of inadequate understanding and/or control of finances and or intellectual property issues.

In principle, the problem is quite straightforward. At all times, you want to have an accurate record of current obligations and receivables, what you owe to others and what others owe you.

When one reads in the press of a substantial enterprise suddenly finding itself in financial difficulties, this is often due to a lack of adequate financial controls. Such unpredictable inability to pay your bills makes potential lenders wary and, at best, raises your cost of borrowing or, at worst, prevents you from borrowing. Since your vendors in effect extend credit from the time they accept your order to the time they deliver, they, too, will require up-front payment and put you into an even worse financial bind.

Since this is presumably understood by everyone in business, how is it possible to fail in this important aspect?

Probably the most common analogy is a man and wife with one checking account and two checkbooks. It is not hard to imagine a situation where some time elapses before they have the opportunity to reconcile their account. If the shortage is large

Table 11.1 Financial plan—base case

Month	Fixed expense	Cum. fixed expense	Parts cost	Cum. parts cost	Tot. cum expense	Boxes shipped	Revenue	Cum. revenue	Revenue—expense
Year 1									
Jan	50	50			50				−50
Feb	50	100			100				−100
Mar	50	150			150				−150
Apr	50	200			200				−200
May	50	250	50	50	300				−300
Jun	50	300	50	100	400				−400
Jul	50	350	50	150	500				−500
Aug	100	450	50	200	650				−650 ←
Sep	100	550	50	250	800	2	200	200	−600
Oct	100	650	50	300	950	2	200	400	−550
Nov	100	750	100	400	1,150	2	200	600	−550
Dec	100	850	100	500	1,350	2	200	800	−550
Year 1 Total	850		500			8	800		
Year 2		850		500				800	
Jan	150	1,000	100	600	1,600	2	200	1,000	−600
Feb	150	1,150	200	800	1,950	4	400	1,400	−550
Mar	150	1,300	200	1,000	2,300	4	400	1,800	−500
Apr	150	1,450	200	1,200	2,650	4	400	2,200	−450
May	150	1,600	200	1,400	3,000	4	400	2,600	−400
Jun	200	1,800	200	1,600	3,400	8	800	3,400	0 ←
Jul	200	2,000	200	1,800	3,800	8	800	4,200	400
Aug	200	2,200	200	2,000	4,200	8	800	5,000	800
Sep	200	2,400	200	2,200	4,600	8	800	5,800	1,200
Oct	200	2,600	200	2,400	5,000	8	800	6,600	1,600
Nov	200	2,800	200	2,600	5,400	8	800	7,400	2,000
Dec	200	3,000	200	2,800	5,800	8	800	8,200	2,400
Year 2 Total	2150		2300			74	7400		

Assume: Parts/Box cost $25k, Box price $100k; Parts costs due upon receipt

with respect to their expected current income and there is little in the way of reserves, multiple checks will bounce, and it may take more than the common grace period to correct the problem. If during this period, they are unable to borrow, this problem will quickly get out of hand. The potential problem is much more serious if family members incur separate installment debt, without prior consultation.

In the typical business, many people are authorized to incur expenses. Ultimately, there is only one cash balance at any point in time. To contain the problem, everyone authorized to incur expenses is given a budget and held periodically accountable. If

Table 11.2 Financial plan—late ship case

Month	Fixed expense	Cum. fixed expense	Parts cost	Cum. parts cost	Tot. cum expense	Boxes shipped	Revenue	Cum. revenue	Revenue—expense
Year 1									
Jan	50	50			50				−50
Feb	50	100			100				−100
Mar	50	150			150				−150
Apr	50	200			200				−200
May	50	250	50	50	300				−300
Jun	50	300	50	100	400				−400
Jul	50	350	50	150	500				−500
Aug	100	450	50	200	650				−650
Sep	100	550	50	250	800				−800
Oct	100	650	50	300	950				−950
Nov	100	750	100	400	1,150				−1,150
Dec	100	850	100	500	1,350	2	200	200	−1,150
Year 1 Total	850		500			2	200		
Year 2		850		500				200	
Jan	150	1,000	100	600	1,600	2	200	400	−1,200
Feb	150	1,150	200	800	1,950	2	200	600	−1,350
Mar	150	1,300	200	1,000	2,300	2	200	800	−1,500
Apr	150	1,450	200	1,200	2,650	2	200	1,000	−1,650 ←
May	150	1,600	200	1,400	3,000	4	400	1,400	−1,600
Jun	200	1,800	200	1,600	3,400	4	400	1,800	−1,600
Jul	200	2,000	200	1,800	3,800	4	400	2,200	−1,600
Aug	200	2,200	200	2,000	4,200	4	400	2,600	−1,600
Sep	200	2,400	200	2,200	4,600	8	800	3,400	−1,200
Oct	200	2,600	200	2,400	5,000	8	800	4,200	−800
Nov	200	2,800	200	2,600	5,400	8	800	5,000	−400
Dec	200	3,000	200	2,800	5,800	8	800	5,800	0 ←
Year 2 Total	2150		2300			56	5600		

Assume: 3 Month slippage in shipping schedule from base case

individual budgets are exceeded or receivables, i.e., amount which your customers owe you, are unexpectedly delayed, only an up-to-date consolidated accounting system can provide adequate forewarning.

The emphasis here must be on "up-to-date." To the experienced financial person, it is the money-time product that is significant.

Tables 11.1 and 11.2 show a sample financial plan designed to illustrate the potential impact of what may appear to be a minor delay in a product program of a start-up company. This company buys parts manufactured by vendors; it does the final assembly and test and sells the finished product, a "box".

To provide a clear overview, only the most essential information is shown. For example, the first column represents costs associated with personnel, i.e., costs which are not directly related to the number of boxes built. These costs would be based upon a detailed "staffing plan" such as shown in Appendix A.3. This staffing plan includes expenses associated with personnel including a percentage of salary for fringe benefits and a 100% burden for overhead such as rent, phone, and maintenance costs.

Similarly, the cost of parts is assumed to be uniform with time. To take advantage of quantity discounts, some of the costs would occur earlier in the cycle with negative impact upon cash flow.

In addition to these "Fixed Costs," there are variable costs here listed as "Parts Cost," associated with boxes in process. The last column, which gives the cumulative revenue minus all of the cumulative expenses, represents the cash flow.

According to the "Base Case" (Table 11.1) a maximum negative cash balance of $650k is incurred in the 8th month, after start-up. Hence, we need to start with a commitment of this amount as a minimum. At the end of 18 months, we expect to be at the break even point. This means, for the first time, we will be operating with funds generated by the business itself. Only if the funds coming in continue to exceed those going out, can we make a profit. If the start-up funds were borrowed, some of these profits go to payoff the debt. Hence "profit" here refers to a positive balance between revenue and expenses, rather than a net profit for the enterprise.

In the "Late Ship Case" (Table 11.2), we have encountered problems in assembly and test. The parts arrived in time (an optimistic assumption) but it took us 3 months longer than planned to get the first box ready to ship. We could not accelerate the shipping schedule beyond that originally planned. We had to pay our vendors upon receipt of the parts.

The complexity of a product designed to sell for $100,000 is such that an unexpected delay of 3 months is not unrealistic.

In this case, our *negative* cash balance has increased by a million dollars, from $650k to $1,650k! Our break even point has moved from 18 months to 2 years. In other words, we need *more than twice* the projected amount of money. If the money is borrowed, we must pay interest for the additional amount and time which, of course, will reduce and delay profitability.

This example was designed to illustrate the consequences of minor delays in a product program. The impact upon cash flow is not intuitively obvious. Nor are such effects readily predicted on the back of an envelope. A going concern will have several product programs going at a given time. This means the allocation of costs to various products becomes an issue. We have simplified this example by not considering the effects of amortization of debt, taxes, and other expenses which are not fixed or directly proportional to the number of boxes shipped, such as advertising or discounts, and allowance for late payment, etc.

> *The intended message*: It takes good financial controls based on good accounting practice to run a business. To ignore this message is to court disaster. To dismiss it as a simple bookkeeping chore is to fail to appreciate the many facets of running a business.

Accounting and financial controls are subjects which can be learned by people who have earned an advanced degree in most other subjects involving quantitative methods. But the learning is not trivial. To repeat: *In a 2 year MBA program you only learn to read the music, not how to conduct the orchestra.*

Similarly, there is no justification to assume that unusually gifted and creative individuals, particularly engineers or scientists, will produce equally good results in financial management. The Nobel Laureate who invented the transistor never did manage to run a profitable semiconductor concern; and it was not that he did not try.

Venture capitalists, the people who provide much of the capital for start-up enterprises, are particularly wary of a team of would-be entrepreneurs that does not include a person with the experience of financial control of a start-up operation.

Eric Ries in his *Lean Startup* (2011) has found a way to articulate these issues which have found wide acceptance.

The following discussion of venture capital (Chap. 12) deals with additional financial aspects of new ventures.

11.5 Summary

This brief treatment of the subject of starting a new enterprise has concentrated upon the essentials:

- The desired personality traits and qualifications of the entrepreneur(s): Ability to make decisions in the face of uncertainty; leadership; salesmanship; and complementary talents among the principals.
- The basis for the new enterprise: Market orientation and access to the targeted customer set.
- The business plan: Emphasis on relevant experience of personnel, market potential, and uniqueness of opportunity.
- The overriding importance of financial controls.

12. Financing Creativity

This chapter points out the role of the various institutions involved in the financing of Science and Hi-Tech. The emphasis is on their impact upon management and upon national policy. The role of Venture Capital, Charitable Foundations, and R&D funding by government agencies is the focus.

(This is an attempt to highlight the importance of this subject and the role it plays in the management of Science and Hi-Tech. It is NOT intended to serve as a "Guide to Project Funding.")

12.1 The Venture Capital Industry

What is the unique contribution of the venture capital industry to the management of creativity? What can the would-be-entrepreneur expect from an association with a venture capital firm? What makes venture capital such a significant force in the Hi-Tech industry?

Of all the aspects of managing creativity in the world of Hi-Tech, none plays a more unique role than the venture capital industry. Its growth over the past 50 years has followed—some would say paced—the emergence of Hi-Tech enterprise.

The Hi-Tech entrepreneur would not have become the folk hero were it not for the synergistic role of venture capital. Neither would the venture capital industry have flourished without the successful entrepreneurs who turned into venture capitalists and applied their experience in order to gain personal leverage.

12.1.1 Who Are the Venture Capitalists?

Many venture capital firms were started by successful entrepreneurs who invested their own funds, realized from public offerings or sale of the Hi-Tech enterprises they had founded. They brought with them both financial and operational experience

R. Kay, *Managing Creativity in Science and Hi-Tech*,
DOI 10.1007/978-3-642-24635-7_12, © Springer-Verlag Berlin Heidelberg 2012

and above all that most elusive of all assets: management know-how in a startup environment.

The growth and financial incentives of the venture capital industry have attracted some of the most talented young people. From the start, the new member of the firm is involved in the decision-making process. It is in the interest of the senior member(s) to impart their knowledge and experience as rapidly as possible in order to increase the investing potential of the firm. This takes the form of on-the-job-training not found in large companies, where top-level decision-making management rarely involves young assistants in anything but staff work.

Venture capital firms may receive 50 or more written proposals in the course of 1 month. Ultimately, 1 or 2% of these proposals are funded. Thus junior members of the venture capital firm, who screen these proposals, may look at as many proposals in a month as their counterparts, new managers in a large established Hi-Tech firm, will see in the first 10 years of their career.

12.1.2   The Venture Capital Network

Venture capital firms compete with each other for funds from investors (mostly institutional investors such as insurance companies, pension funds, and major endowments and foundations). On the other hand, the disbursement of these funds is a well-coordinated, cooperative effort among several venture capital firms. Given the very large number of proposals which are submitted to the venture capital firms, no one firm could do the detailed analyses and investigations necessary to support many multimillion dollar investment decisions. What is more, rarely does a venture capital firm become the sole investor in a new venture or in subsequent rounds of financing. Generally, the number of venture capital firms participating in a particular venture varies between 1 and 10, depending upon the amount of money involved.

The cooperation among the venture capital firms is informal. The principals of established firms frequently know each other personally. A firm will take the lead in financing a startup venture based upon unique and relevant experience. It does so with the understanding that it thereby takes a responsibility on behalf of the other venture capital firms which will participate in such a syndicated effort. This responsibility is not taken lightly; each firm is dependent upon others for inclusion in future syndicates.

The benefit of syndication extends beyond the sharing of the considerable effort involved in the evaluation of proposals. In fact, its primary purpose is to spread the risk (associated with a given venture) among several venture capital firms.

12.1.3   The Risk and Reward of Venture Capital

The stakes are high, and so is the risk. The investment of the venture capital firm's own money usually represents one or two percent of a given investment fund. In addition to an annual management fee of 2–3% of the fund total, the venture capital firm will realize 20% of the ultimate gain or loss when the fund or a part of a fund is cashed out. A given fund, typically, represents a 10–100 million dollar investment pool.

Thus, the venture capital firm's ultimate investment performance and ability to attract investment capital are based upon the financial performance of the portfolio companies during the life of the fund, generally 7–12 years.

What makes a startup enterprise attractive is the fact that it does not need an existing large customer base, only the potential thereof. This means that it can prove itself with relatively small initial commitments. The established industry leader, on the other hand, is focused on maintaining market share by incremental improvement of its present product line(s) serving a large customer base and broadening that base.

12.1.4   Accessibility of Experienced Management

The venture capital firm which takes the lead in financing a particular venture, be it a start-up or a subsequent round of financing or taking a company public, assumes responsibility beyond raising capital. The larger venture capital firms pride themselves in nurturing the young venture.

This nurturing takes the form of management counseling and, at times, management support; executive recruitment, contacts with important customers or vendors, and occasional active intervention if the occasion demands. When the situation warrants drastic action, a senior partner of a venture capital firm may even assume the temporary chairmanship of a troubled company. These relatively small companies thus have, at least for a crucial period of time, the benefit of some very experienced top-management talent. This is not an act of altruism on the part of the venture capital firm. Its own performance is influenced not only by market conditions in general, but also by the relative performance of their portfolio companies.

The venture capital firms are always represented on the new venture's board of directors to look after the interests of the investors; under ordinary circumstances, they will not play an active management role.

While the interaction between the management of the new venture and the capitalists is not without friction, the common objective tends to promote understanding. It is the objective of venture capital firms to invest large sums of money and to protect that investment, not to manage new ventures. Hence, direct intervention is only a last resort.

Hi-Tech entrepreneurs who have been successful without a venture capital firm almost invariably have had significant management experience, which has included marketing and financial responsibility. Also, they must have had access to sources of investment capital.

While there is considerable folklore about the success of people without management experience, much depends upon the definition of "success." The decision to *share control* of a successful enterprise is generally preferable to the *sole control* of a failing enterprise. What is more, initially successful entrepreneurs have been known to lose control and even be forced out of a going enterprise when their lack of management experience catches up with them.

12.1.5   The Related Infrastructure

Another factor in the success of the venture capital industry is the evolution of an infrastructure of related organizations which play an important role in this industry. Numerous corporate law firms specializing in startup ventures, venture capital financing, and public offerings provide valuable counsel to the new venture. Executive recruiting firms, law firms specializing in patent law, and experienced accounting firms are all part of the informal network which ties the venture capital industry together. The geographic clustering of the venture capital industry in California, Massachusetts, and New York has minimized the problem of critical mass, and it has been possible for these firms to maintain the personal contacts which are essential for such an informal network to function.

An example of a startup experience will bring this complex arrangement into focus.

Three engineers in the XYZ Corporation which itself was started 5 years ago have approached XYZ's management with a proposal which was outside the current product scope. At the time, XYZ itself was contemplating additional financing for expansion of their existing business based on the current product line.

The reason for turning down this proposal was based upon the judgment that available financial and management resources of XYZ would be better devoted to existing opportunities associated with current products. This view was confirmed by the venture capital firms which had provided earlier financing for XYZ.

The three engineers take their proposal to an acquaintance who is associated with a local law firm specializing in venture capital start-ups. In addition to a critical review of the proposal, the lawyer calls up a personal contact in a venture capital firm. The lawyer briefs the entrepreneurs on the firm they are about to contact and introduces them with personal recommendations. It turns out that the contacted firm is unable to assume the role of lead investor because of other commitments. They facilitate contact with another venture capital firm, having seen merit in the proposal and can assume that they will be invited to participate in a financing syndicate.

Once the venture capital firm has established interest, a financial arrangement between the three entrepreneurs and the venture capital firm is negotiated. Both parties, the startup management and the venture capital firm, are represented by lawyers experienced in startup ventures.

The entrepreneurs' desire to maintain maximum control contends with the venture capitalists' assessment of the potential for ultimate success and the impact upon future rounds of financing. The venture capitalists are influenced by the personal risk the entrepreneurs are prepared to take: how much of their own resources they are prepared to commit. The venture capital firm, of course, strives to improve its own performance; hence, past arrangements are a guide.

If these negotiations do not lead to a mutually satisfactory arrangement, the entrepreneurs can seek out another venture capital firm. Since there are usually strong personalities involved on both sides, there may be a good rationale for selecting one venture capital firm over another, based purely upon the people involved. By the nature of the venture capital industry, the odds are not with the entrepreneur to negotiate a significantly better deal.

Given a satisfactory basis for a financial arrangement, that is, distribution of equity in the new venture between the three engineers and the venture capital firm(s), the lead venture capital firm makes a prompt and fairly detailed investigation of every aspect of the proposal to meet the requirements of "due diligence" defined by the regulatory agencies that oversee the investment community.

These requirements are significantly more stringent when an existing privately financed venture is to be taken public. The due diligence process culminates in the drawing up of papers of incorporation, involving lawyers who specialize in this aspect of corporate law. Our entrepreneurs have interviewed several such law firms recommended by their lawyer friend and by the venture capital firm.

The due diligence process is facilitated by the existing network among the venture capital firms. Someone in one of the firms in the network will have personal access to the entrepreneurs' former employer, colleagues, or references.

The above scenario of a startup venture was designed to illustrate the workings of the infrastructure. It does not take into account all aspects of the process. The inexperienced entrepreneur seeking financing generally will have to try several sources of potential financing. Even with seasoned advice, this can take 6–12 months. (Do not be misled by the startling claims of the outliers.)

12.1.6   Evaluation of Proposals

In the previous discussion of the evaluation of R&D, particularly the evaluation of proposals for new projects, we have pointed out the difficulty of this many-facetted task. We have also indicated the importance of the business plan to the startup enterprise.

The venture capital industry has a unique and most effective method of evaluating a business plan, after the initial screening has determined that it meets the current criteria of interest of a particular VC firm. Executives of the portfolio companies of a venture capital firm, that is, the companies in which the venture capital firm has already invested, constitute a unique pool of "consultants." Among them will be entrepreneurs with the technical, financial, market, and management experience relevant to the proposal under consideration. Often the company executives consulted will have closely related interests which could be affected. Hence, the quality and insight which are brought to bear upon the evaluation are uniquely relevant.

The consulted executive is part of the infrastructure which is based upon mutual dependence: the consulting executive needs the VC firm for future capitalization and the VC firm is dependent upon his or her expert advice.

This is in stark contrast to the competitive environment and "Not Invented Here" (NIH) syndrome which lines the rocky road to be traveled by the potential innovator within most established Hi-Tech companies. The relative speed with which an interesting proposal can be acted upon by the venture capital community is one of the motivations for the entrepreneur-to-be to consider starting a new venture. Frequently, he or she has been frustrated by the fact that in the established company, accessible levels of management, concentrating on their own task, look upon a new idea as potential competition for limited resources.

Even when an innovative proposal falls upon receptive ears, the process of evaluation and integration into the "planning cycle" of an established corporation generally is not as responsive as the venture process which is uniquely geared to "new opportunities."

The criteria used by the venture capital firm to evaluate the proposal, in order of importance, are as follows:

1. *Financial performance*: The venture capital firm is looking for ventures with the prospect of earning a compound rate of return on investment at least 20% greater than relatively risk-free alternatives, such as secured loans or Treasury notes. In this calculation, the time horizon is as significant as the capital requirement itself, because investment decisions consider the present value of dollars earned in the future. At the generally used discount rate of 15%, for example, the present value of a dollar earned 5 years from now is 49.7 cents; in 10 years, 24.7 cents; and in 20 years, 6.1 cents. This present-value method makes it apparent why maintaining the projected growth pattern in order to go public, be acquired, or merge within 5–7 years, is critical to success.

2. *Management experience* of the founding team: Even the best proposal is only realizable to the extent that management has the ability to implement the business plan. Assume that the venture capital firm likes the idea that is the basis for the product or service upon which the venture rests, but recognizes a weakness in the proposed management team. The VC firm may stipulate the inclusion of a chief executive officer with startup experience in the founding management team.

3. *The uniqueness and marketability* of the product or service to be offered by the new venture: There is a substantial body of knowledge, mostly in the heads of experienced venture capitalists, which is used to assess that aspect of the proposal. Some of the issues are the following: Is the offering essential to the customer's business and will it enhance the customer's productivity? A large customer who is totally dependent upon such an offering may not want to be captive to a small startup venture. What is the projected market share of the new venture? Experience has shown success to be dependent on a more than minimal market share.

4. The technical aspects of the proposal are important but rarely a matter of contention, even in Hi-Tech undertakings. It is generally less difficult to get the consensus of experts about technical feasibility than about marketability.

Since relevant knowledge and contacts are specific to each type of business, a venture capital firm will generally confine its activity as lead investor to those areas where it has access to branch-specific knowledge, experience, and business contacts.

12.1.7 The Relative Size of Venture Capital Investment

The bulk of the R&D expenditures of established Hi-Tech firms goes into the enhancement and extension of current product lines and proven technologies. An estimated 10% goes into long-term or basic research and what is sometimes referred to as "advanced technology." Such advanced technology would include new products and services which do not address established market needs or which address new, unproven technology.

The venture capital industry, on the other hand, invests primarily in new products and services and commercially unproven technology.

The rate of Venture Capital investment depends largely on the prevailing investment climate and consequently fluctuates on an annual basis.

Available data would suggest that venture capital could account for as much as one half of the new product and new service investment associated with Hi-Tech initiatives. Thus, while venture capital accounts for a relatively small percentage of the total R&D expenditures, particularly when defense-related R&D is included, it has served to direct significant investment capital to the financing of *new* Hi-Tech ventures and thus has had a major impact upon innovative effort.

It is important to note that while venture capital has played a significant role in the growth of high technology, its aim is to serve large investors whose focus is on expected return, not on the needs of high technology.

This means that attention and investment capital can shift readily to a more promising opportunity in a totally different field, e.g., from information technology to biotechnology or from the Internet to alternative energy sources. The uniqueness of the venture capital industry resides in its ability to address a wide variety of needs involving diverse technologies, once a critical-mass VC infrastructure is in place.

As a relative newcomer among financial institutions, the venture capital industry has upheld high standards. Yet, major downturns in the economy do not spare the venture capital industry. Low or negative ROI puts pressure on the industry which exists by virtue of its promise of high returns. This tends to place undue emphasis upon shorter term returns, i.e., puts the new start-up at a disadvantage in times of economic downturns (Zaytsev 2011).

By and large, the VC industry has been effective in channeling investment capital from established financial institutions and wealthy individuals into the hands of entrepreneurs, who have thus been able to create the industries of the future.

12.1.8   Summary: Venture Capital

The venture capital industry has played an important and unique role not only in the financing but also in the management of creativity during the past 50 years.

A valuable attribute of the venture capital industry is the unique experience it can bring to the start-up of a new enterprise. The principals of many venture capital firms are former entrepreneurs.

Syndicated financing is the basis for collaboration and serves to spread the risk among venture capital firms. A network which includes the VC portfolio firms enhances this collaboration, as does an infrastructure of supporting organizations which provide legal, accounting, and recruiting services.

Venture capital firms can offer one of the most desirable assets to a new venture, i.e., access to relevant management experience.

The risks and rewards inherent in the venture capital industry establish the basis for investment criteria, in particular, the element of time.

The venture capital industry's criteria for the evaluation of a prospective new venture focus on growth potential, the entrepreneur's relevant experience, and the marketability and uniqueness of the product or service.

(The relative emphasis placed upon the venture capital industry is based on the conviction that it represents a unique aspect of managing creativity. It emerged from the cradle of high-tech, founded by some of the most capable innovators whose entrepreneurial bent added a rare talent to their scientific and technical achievements. They have demonstrated a way of leveraging managerial talent.)

(See Chap. 14 for some international aspects of Venture Capital.)

12.2   The Role of Foundations

Limiting our view to Charitable Foundations focused upon Hi-Tech and the underlying scientific disciplines, they play another unique part in the financing of creative effort. On the one hand, they fill a need created by the successes of capitalism: They provide a vehicle for investing large personal fortunes in efforts to better the world. On the other hand, foundations are able to support groundbreaking efforts which are generally outside the compass of government or the private sector.

Charitable foundations are essentially American institutions; only in recent times have they spread to other parts of the world. They generally represent the vision (or idiosyncrasies) of their founders and surviving family members. Over time, foundations evolve into organizations led by people who have demonstrated outstanding capabilities in some endeavor related to the activities of the foundation. They are attracted to such positions by the prospect of directing significant resources to new and worthwhile endeavors which promise to have a positive impact on society. The uniqueness of their position is their relative independence; they generally define and select the objectives and specific programs to be supported.

While the foundation trustees have the power of terminating a director, they usually are not involved in the selection of programs.

What foundations have in common with the venture capital industry is their focus upon start-up rather than sustaining efforts. This is based on the premise that once the societal or commercial promise of an idea has been demonstrated, public or private organizations will be motivated to carry on.

Missing from the structure of foundation support, compared to venture capital, is the ability to influence a project beyond funding. The foundations are generally not equipped to mentor a project.

A unique feature of foundation support is that it is not restricted by the constraints of most established organizations, i.e., profitability or peer review.

The positive aspects of peer review are well understood: Proposals or ideas based upon an inadequate understanding of current knowledge can be filtered out; the relative merits of competing proposals can be evaluated.

The negative aspects of peer review are not as obvious. Peer groups can be subject to undue influence by financial and political interests, not to speak of the peer group's interest in self-perpetuation, i.e., protecting its turf.

To illustrate the implication of this, imagine two competing proposals to study the diagnoses or treatment of cancer: One based on infrared radiation which would replace current skills and equipment with less costly treatment and equipment, and the other aimed at evaluating the promise of extending the range of applicability of current methodology.

In the USA alone, at stake are the interests of 30,000 radiologists and the folks responsible for their 6 years of post MD education and a 6.6×10^9 medical imaging equipment industry.

While profitability, per se, does not influence the way a foundation allocates its resources, there are financial considerations which matter.

From the point of view of the philanthropist, a one-time gift to a worthwhile nonprofit organization may loose its luster over time, while a foundation can control its future relevance.

Charitable foundations owe their existence to favorable tax treatment; in effect, a significant portion of their funding is taxpayer supported. (An outright gift to a qualifying nonprofit organization produces a one-time tax deduction based on the amount of the gift, an amount no longer under the control of the philanthropist.)

A foundation affords a deduction based on the total value of the foundation, but only 5% of that value need be disbursed per annum. The "Foundation" retains control in perpetuity. Given the fact that collectively, Foundations are valued at several hundred billion dollars, their modus operandi may be challenged in time (Porter and Kramer 1999).

These particular strength and limitations of Foundations result in a modus operandi which is opportunity rather than strategy driven.

Given the fact that most Foundations have been able to attract very capable people and can point to many exceptional achievements, their overall contribution has gone unchallenged.

12.3   Government Sponsored R&D

Again limiting our view to Science- and Hi-Tech-related effort, governments perform a very important, often decisive, role. In most industrial nations, government initiative and support have been key factors in Science and Hi-Tech.

While "national defense" plays a major role, many national governments have instigated or supported initiatives to advance Science and Hi-Tech as a means to attain or maintain competitiveness. This may take the form of government operated or fully supported laboratories as well as funding of universities and industries and of various independent institutes or laboratories.

Many such independent institutes are associated with universities and are supported by government-funded grants which are awarded in competition with university groups pursuing similar objectives. They are in part supported by a wide variety of nonprofit organizations and include industrial consortia.

Grant proposals are usually subject to peer review and can be designed to promote collaboration, to encourage interdisciplinary efforts, or to undertake efforts beyond the scope of a single group. By awarding such grants to multiple independent groups, the granting agency also avoids some of the constraints imposed by peer review.

The European Union has used such collaborative grants to help member countries to catch up in a particular field of science or technology. More recently through its Flagship Program, EU support has tried to provide resources to match similar efforts in other parts of the world.

Authoritarian regimes have used the control of R&D funding to promote a science and technology policy which supports political or religious goals.

It is evident that the role of government goes well beyond its impact upon the management of Science and Hi-Tech. It is a large topic which we cannot do full justice here. It is but another factor in the growing importance of the role of management of creative effort.

12.4   Conclusions

The analysis presented above is suggestive of far-reaching implications for any National Policy for R&D funding.

In the USA, Venture Capital and Charitable Foundations have played a significant role in the burgeoning of Science and Hi-Tech in the twentieth century.

As a consequence, the role of Venture Capital and Charitable Foundations is growing in countries aspiring to play a role in Science and Hi-Tech.

In Europe, diversity of funding derives from their sources being controlled by European Union or National agencies.

Government policy toward Science and Hi-tech will continue to have a decisive impact.

The management of nonprofit R&D institutes is critically dependent upon a far-reaching understanding of the workings of the varied sources of government funding available to the pursuit of Science and Hi-Tech. This has become an increasingly important issue the world over.

This brief review raises the question: What is it about this combination of institutional approaches that accounts for its evident success?

- Government and industry in-house R&D, the largest components of funding, address major national needs and capabilities.
- Venture capital is uniquely capable of taking on high-risk efforts.
- Foundations are not restricted by national interest or profit potential and can depend upon the public and private sector to pursue proven ideas.
- Government sponsored R&D can address long-term and interdisciplinary objectives directed at regional and local interests (i.e., EU, national, state, and local).

It is the organizational design of each of these institutions which defines their unique strength and limitations.

National governments must balance the needs of security and economic development; priorities are set by political considerations. [1]

Industry must seek a balance between short-term and long-term profitability. In defining priorities, it is subject to the influence of past success, i.e., the basis of their current revenue.

Venture capital has necessarily a short-term perspective, driven by profitability. There is no long-term commitment to a particular industry or national interest. They have an unmatched task-relevant ability to exercise managerial control when warranted.

Foundations are not constrained by self-serving needs. Neither are they subject to the limitation of the peer review system. They can depend upon public and private interests to pick up and sustain their successful programs.

Government grant sponsored R&D can address and sustain major long-term initiatives. It can be subject to political interests and the limitations of peer review.

It is difficult to conceive of an organizational construct which would incorporate all of the capabilities without any of the limitations of the varied institutions which have gained acceptance and have had such remarkable success. [2]

---

[1] In the European Union, the resources are controlled by local, regional and national governments. This results in diverse priorities, designed to serve a wide range of interests.

[2] The largely successful effort on the part of Germany in the 1960's to jump-start a high tech effort through a broad and far-reaching government program could not be sustained without an effective Venture Capital Industry required to turn such R & D into innovative products and services (Thomas 2005).

12.5   Summary: Financing Creativity

Industry and government provide the major support for Science and Hi-Tech efforts; their priorities are dictated by industry-specific and national needs, respectively.

The venture capital industry addresses the unique needs of innovative effort outside the scope of established organizations. These efforts are motivated by high return on investment with commensurate high risk. In addition, venture capital can provide relevant management experience to the new venture.

Foundations are in the unique position of supporting efforts designed to solve or impact societal problems, without the constraint of profitability or national interest. They enjoy the rare position of being able to direct their resources to worthy efforts, so defined by themselves. They are free to seek input from the most highly regarded experts, but are not beholden to peer review.

Government support through grants to universities, industry, and independent nonprofit institutes can provide long-term support including interdisciplinary efforts and involving multiple diverse institutions.

Collectively, such varied sources of funding make possible the pursuit of varied strategic objectives. (Chapter 14 touches upon some international aspects of this topic.)

13. Organizational Culture

Throughout this book, it has been maintained that there is no single approach to problems confronting a manager. This may have given rise to occasional ambivalence—a feeling all too familiar to the conscientious manager. Conveying this element of ambivalence is an effort to encourage you, the reader, to recognize the need for personal choice in dealing with such management problems. This chapter identifies some of the principles which can help in the approach to such personal choice.

A combination of characteristics makes high-technology organizations unique. These characteristics include high rate of change, important role of creativity, acceptance of high risk, and the need to rely upon unproven approaches.

To succeed in such an environment, it is desirable to maintain flexibility and to keep formal procedures to a minimum. Without such procedures or when they are not applicable, the organization needs a set of principles based upon a common set of values, *an organizational culture*, to provide guidance.

The basis for such cultural values is the precedence of the interests of the organization over those of a component group or of an individual. That is to say, everyone needs to understand and accept values and principles which apply to everyone in the organization. Loyalty to an organization is a derivative of an individual's acceptance of such values. Acceptance requires that these values be articulated explicitly. How can this be done?

Top management must find and realize every opportunity to elucidate its corporate culture, be it in one-on-one or group meetings, and be it through in-house publications, but above all by constant example.

Some may question the need for cultural values altogether. It is important to realize that organizational cultures are a given, even if they have come into being without conscious intent. What is more, some aspects of corporate culture can have negative, even destructive ramifications; probably the most insidious and difficult to obviate—the "arrogance of success" (Gerstner 2002, p 109).

How does one recognize the elements of an organizational culture? Steve Brandt has found it efficacious to define—or better said, recognize—the culture

R. Kay, *Managing Creativity in Science and Hi-Tech*,
DOI 10.1007/978-3-642-24635-7_13, © Springer-Verlag Berlin Heidelberg 2012

of an organization by looking critically at six factors borrowed from anthropology (Brandt 1986):

1. Vocabulary

"Words (and phrases) are the tracks that ideas and values run on. Without a word for it, an idea or concept cannot exist in a company setting. Without the words, 'quality' or 'entrepreneurship,' it is difficult to get more of either, for example."

2. Methodology

By methodology, Brandt means the established ways in which an organization gets things done, including the use of technology, for example, internal task forces or external consultants and teleconferencing or meetings in a beach resort.

3. Rules of Conduct

The unwritten dos and don'ts that guide day-to-day actions and behavior: From the dress considered appropriate to various situations and office protocol to decision making and etiquette. There are organizations where an outsider attending a meeting has no immediate way to identify the person in charge—others where position at the table, special chair, and the ratio of talking to listening leave no doubt.

4. Values

In successful enterprises, positive values are in evidence. Among these values, the authors of *In Search of Excellence* (Peters et al. 1982) identified:

- A belief in being the "best."
- A belief in the importance of people as individuals.
- A belief in superior quality and service.

5. Rituals

Upon arrival, Professor Smith, a visitor, finds a prominent sign in the lobby "Welcome Prof. A. Smith, Dept. of Mathematics, University of Edinburgh." The lab director, rather than a secretary, comes to meet Prof. Smith in the lobby and escorts her.

There are more universal rituals such as various types of announcements to the membership at large, holiday parties and the way new people are introduced, all of which reflect the values of the organization.

6. Myths and Stories

Who are the heroes? Who is the butt of the jokes? What do the people from the lab talk about after a few beers together? Are the stars acknowledged to be those making positive contributions or are they the people beating the system? Is it good or foolish to work long and hard?

13.1   The Matter of Trust

It is not easy to discuss, let alone write about concepts such as trust, loyalty, tolerance, and personal integrity without preaching or sounding presumptuous. We will try to overcome this difficulty by presenting a number of examples.

The most important expectation which derives from a common set of values is the development of trust. Given a common set of values, people and situations become more predictable. Lack of predictability leads to distrust. One of the most painful lessons is to realize the loss of someone's trust and to find that the task of reestablishing trust is exceedingly difficult.

Loyalty to another person or to a group is conditional upon the loyal individual having the trust of the group or other person. It is difficult to imagine being loyal to a group or organization which gives every indication of not trusting you. In the following examples, the issue is the role of trust in the resolution of potential conflict of interests. It should help put the concept of trust into concrete terms.

A husband and wife work for competing companies. One has a responsible management position, and the other is being considered for such a position.

In a similar situation, a person in a responsible position is about to marry a person in a responsible position in a competing firm.

In the first case it may be argued that the company has the option of not promoting the spouse, thus avoiding possible exposure. In the second case, neither company wants to lose a valuable employee.

There are some organizations where the cultural values are quite explicit in placing absolute trust in the employees. Where this is the case, the relationship to someone in a competitive firm cannot be the basis for judging trustworthiness. This absolute trust is articulated explicitly at the time of recruitment, and periodically thereafter. That is, all employees are made aware of their responsibility to identify and reconcile possible conflict-of-interest situations.

The explicit statement may be: "Should the matter of conflict of interests ever come up, you are obligated, as a condition of your continued employment, to discuss the matter with your manager without delay."

Another organization, a small company in a very competitive business, may not be willing to accept the risk and opt for a policy of making a case-by-case decision about continuing employment in the case of potential conflict of interests. This would imply that the company, not the individual, decides upon the manner in which "potential" conflict is defined and reconciled.

The values upon which such policies are based provide a clear choice. In either case, the principles which are upheld can serve the organization only to the extent that they are understood and accepted by all involved. It should be added that without consistent and uniform application of such policies, the essential element of predictability is missing. Another example:

Negotiations concerning a contractual arrangement failed when it became apparent that Erwin did not trust me. Subsequently, the reason for Erwin's lack of trust became clear. Had he been in my shoes, he would not have hesitated to go back on his commitment, had a more advantageous deal come up after our agreement. Since he had little or no basis to judge my trustworthiness, his lack of trust was a reflection of his own set of values.

While the matter of trust dominates all interpersonal relations, debates about intellectual property assume special significance for creative people in the Hi-Tech environments, as we discuss now.

13.2 Intellectual Property: Whose?

One principle that has proven particularly relevant to managers of creative people— Don't compete with the people you manage.

The practice of putting the manager's name on papers representing the work of people in the manager's group, when the manager did not make a unique technical contribution to the reported work, is poor management judgment. When there is room for doubt the manager must back down; otherwise, he/she may create the impression that managerial authority, rather than the objective merit of the issue, is the basis for the decisions.

It is difficult to avoid disagreement about intellectual property: Whose idea was it? Who thought of it first? Where is the line between making a suggestion and making a contribution? What value attaches to the "definition" of the problem? The manager is called upon to arbitrate such issues. This is very difficult when the manager is thought to have taken undue credit, i.e., laid claim to the property of others. You, the manager, have to determine to what extent the perception of others will impede your effectiveness. Indeed, it is the perception of others, not the righteousness of your position, which determines that effectiveness.

Some creative people who have gained stature by virtue of their publications or patents find this principle difficult to accept. For example:

The director of an industrial research laboratory enjoyed the reputation of being the most prolific inventor in the company. His name was included in every patent application filed by the laboratory. This led to the filing of a lot of worthless patents, since his implicit judgment about the subject of the application was beyond challenge. It was not until some years after the retirement of this director that a meaningful set of criteria was defined for internal evaluation of patent applications in this organization.

The practice of listing joint authors in alphabetical order is preferable to that of putting the senior person or manager first. The best reason for including the name of someone well established in the field is to afford name recognition and the implied, associated standard of excellence. In a university, to give top billing to the person who has done most of the work reflects a value system which assumes the professor's role as that of providing opportunities for the recognition of their students, rather than the other way around. Competition among academic institutions and the fact that the best among recent graduates can find opportunities even beyond national boundaries have served to reduce abusive practices which were long defended on the basis of tradition.

13.3 Expense Accounts: Judgment or Integrity?

The handling of expense accounts is another frequent source of contention when the basic principles at stake are not understood clearly by everyone. Two specific examples come to mind:

Sam is a highly valued scientist, respected for his outstanding contributions. He repeatedly includes items in his expense accounts which are clearly not legitimate: Cost of a dinner included in the conference registration already paid for by his organization; personal travel expenses incurred over a weekend, while on a business trip. There is nothing deceptive about the way these expenses are stated. Sam believes that being the outstanding contributor he is entitles him to such "extras."

Bill submits an expense account which includes an exorbitant gourmet dinner. He defends this expense as appropriate since it was his birthday away from home. He is accustomed to treat himself to such a dinner on every birthday.

Accepted accounting practice does not allow such items as lawful business expense. There should be no problem making this clear to all.

A more difficult issue arises when the "padding" occurs within the defined acceptable limits or there is some basis for suspecting a fraudulent claim. The principles at stake here are as follows:

- Expense accounts are not exempted from the expectation of mature, responsible judgment.
- A mature, responsible individual would not willingly allow his or her integrity or judgment to be suspect.

There is a fairly straightforward way of dealing with such problems.

1. Make sure that there are clear guidelines defining the acceptable limits and that everyone is aware of them.
2. When an exception is thought to be warranted, have the justification submitted with the expense account.
3. Any item which fails to meet the guidelines or does not justify an exception is disallowed.
4. Proven instances of deliberate misrepresentation are looked upon as theft of property, which makes the individual subject to immediate dismissal.

Without such a procedure, the manager may be put in the situation of repeatedly questioning a subordinate's judgment and/or integrity. Unwilling to do this, the manager may choose to ignore the matter with the result that a clerical person in the accounting department must escalate the issue thus involving other levels of management.

To ignore such issues altogether will lead some to conclude that responsible, personal judgment and integrity are not part of the cultural values of the organization. Expense accounts generally involve relatively small sums of money. This makes them a preferred vehicle to explicate these particular values.

13.4   Bureaucracy: A Positive Asset?

Scientists and engineers see "bureaucracy" as the very antithesis of creativity. Established procedures and channels of approval are looked upon as creativity-stifling "red tape."

As consultant to Hi-Tech organizations, I have encountered frequent instances of "technical problems" which, upon closer inspection, turned out to be management problems. Very often, these problems could have been avoided, had there been well-established and documented guidelines which had the backing of top management. Again, some examples:

A complex and interrelated set of problems in transferring a product from development to manufacturing turned out to be caused by a lack of clear assignment of responsibility. When the new product to be transferred turned out to involve technology new to the manufacturing people, recriminations about "incompetence," on both sides, became the order of the day.

A mutually agreed upon schedule, identifying responsibility and a procedure for resolving conflicts, was the obvious way to deal with the problem. Earlier attempts to establish such procedures had been rejected as unwanted bureaucracy.

The CEO of a Hi-Tech company with annual revenues exceeding $400 million, at some point, found several of his key people inadequate to do their task. It turned out that the CEO himself had hired these people over a period of years. His "selection procedure" was essentially based on his "first impression"; he had great faith in his intuition. Having come through the sales organization, he applied criteria that were inappropriate to scientists and engineers in a Hi-Tech R&D environment. He had hired the people reporting to him directly, without involving his personnel department, which he referred to as a "bunch of bureaucrats."

An otherwise successful start-up company had an excessive turnover of key employees. Excellent working conditions and challenging work made this surprising. After interviewing several of the people who had left, it was found that the CEO had made financial arrangements with each of these people which they did not consider equitable. In place of some consistent and equitable basis for granting stock options, etc., he chose the principle of "the least I can get away with." The CEO had not expected this information to be shared. By taking this position, he had undermined trust and forfeited the kind of feedback which could have corrected this situation.

In a well-established organization, it is feasible to have a new employee sign a very simple employment contract. This is possible since all of the underlying assumptions about vacations, time-off, salary review, various benefit programs, conditions of termination, intellectual property rights, etc., are well defined and documented. A newly established organization, or one which rejects the notion of a "book of procedures," has a choice: ignore the responsibility to protect itself and the employee, or accept a fairly lengthy and complex contract which may well scare away some prospective employees who fail to see this as "less bureaucracy."

A recently hired engineer, considered to be an outstanding individual by everyone in the development laboratory of a major corporation, resigned out of frustration with the lack of freedom in making decisions. He joined a small company which offered "complete freedom" to decide what to do and how to do it. It took only a few weeks for this engineer to realize that, since everyone else had joined with the same expectations, the company was pursuing more projects than it had resources to carry to fruition.

A research laboratory administration has proposed to save manpower by eliminating mail delivery to individual offices and instead provide departmental mailboxes. The affected researchers accepted this suggestion on the condition that the manpower resources saved would be allocated to the scientific staff. The rationale: A Ph.D. chemist can pick up his or her own mail, but you cannot expect a person hired to deliver mail to do research. Since the total resources of the organization are limited, increasing the scientific staff was held to be more effective in pursuing the goals.

Lest there is any doubt: I am against bureaucracy per se.

I would also reject any attempt of administrators to introduce "procedures" or "rules" which are justified by streamlining administration or making the job of administration easier; particularly when this is to be done by shifting the burden onto the scientists and engineers upon whom the organization depends for results. In a Hi-Tech environment where creativity is to be nurtured, introduction or change of any administrative procedure should be subject to line management concurrence.

We should be even more adamant in our rejection of procedures that are defended purely on the basis of their long-lived existence. Creativity can flourish only in an environment which allows for change.

There are some organizational entities that are constraint to operate with rules and regulations that impede creative endeavor. They may be part of a larger organization such as a government agency, or a large corporation. They may be subject to organizational cultures that make it difficult to effect turnover of personnel or alternative career paths. Some national cultures put age above relevant ability. Bringing about change in such an environment is often very difficult. Organizations that are unable to effect change must accept the fact that in the world of high-technology they will be at a distinct competitive disadvantage.

Creativity will diminish when concern with anything but results stands in the way of change.

13.5 Cultural Values

The above examples are intended to relate such abstract concepts as trust and integrity, which we have called *cultural values*, to the concrete experience of scientists and engineers. The occasions to focus upon these values in a situation-relevant fashion are rare. It is a subject generally neglected by our educational system and by most available formal management training. Yet, experienced managers would agree that a good portion of "real" management problems relate to this subject.

While I do not feel called upon to advocate a particular set of values, there are some guidelines for applying such values.

It is essential that "success" and the values which the organization expects to uphold in the pursuit of success are made explicit. A lack of positive values is perceived as endorsement of negative values.

Promulgation of such values must be in words and deeds. To be generally accepted, the upholding of such values must be in evidence at all times. When procedures designed to uphold these values are not applied uniformly or consistently, their effect will be counterproductive.

Where do these values come from? There is no question that the people ultimately responsible for the success or failure of an organization must define and identify with a set of values which are to serve the organization. They not only must be counted upon to uphold these values every time they are challenged, but they must be diligent in the periodic restatement of these values, lest someone questions their continued relevance.

One of the largest State universities in the United States, to maintain its position as one of the leading academic institutions in the world, holds "Excellence" as the value of uppermost importance. Every new faculty appointment is cause for reiteration of this position. Many arguments are advanced, based upon internal and external political considerations, to argue for a candidate who does not meet the criterion of excellence as well as someone else. To back down on the primacy of this value would reduce the university's standing over the long run. But more importantly, to back down would expose every faculty member to pressure to select students or give grades on some basis other than achievement. By the same token, this university does not award honorary degrees, undoubtedly at some cost in terms of potential donations.

A Hi-Tech enterprise which operates successfully in many countries of the world has consistently avoided scandal by placing high value on personal integrity. This means rigorous enforcement of rules which leave no doubt about the level of tolerance toward lacking integrity. They regard being a little dishonest as one would regard being a little bit pregnant. As a consequence, several hundred thousand employees of this organization enjoy great respect and superior credit ratings in the communities in which they live.

A relatively new and small company providing products used by professionals, software for personal computers, has built its success upon the notion of providing value to their customers. This is a guiding principle. "Value to the customer" is here defined not only in terms of price and quality, but also in terms of understanding the needs of their customers. Many ad hoc decisions in this loosely run organization are influenced by the value attached to this goal. It is difficult to imagine a more relevant goal in such a very competitive environment.

13.6 Respect for the Individual

Respect for the individual is that part of the value system which tops the list in dealing with people. Many of the unexpected crises for which no one is prepared— at work, at play, or in the home—can often be resolved in a satisfactory way if the guiding principle in making decisions is respect for the individual.

This respect for the individual, its existence or lack thereof, is readily evident to the sensitive observer: The attitude of the receptionist or the person answering the phone, the general appearance of people and facilities, the treatment accorded to secretaries, the listening skill of the peer group, and the perceived role of the manager.

The management of a restaurant, through its staff, creates an ambience which reflects a set of interpersonal values. It is all too obvious, when this ambience is at either end of the spectrum.

More often than not, your attention is called to such manifestations of organizational culture by something perceived negatively. If the manager is rude to his or her spouse because of annoyance at an interruption by the spouse's phone call, the inadvertent listener will go away with a feeling that this manager is not likely to carry the professed respect for the individual into practice. Lack of consideration for the person closest to the manager cannot fail to make an impression. If the top person in an organization appears to treat a secretary with less respect or consideration than other members of the staff, it is most likely that this attitude will permeate the organization. Differential respect, once established, can easily affect others, particularly junior members, minorities, etc.

This is not to imply an expectation of uniform attitudes to such personal matters as how you communicate with your spouse. Nor is the objective the acceptance of a particular uniform pattern of behavior.

These examples aim to demonstrate the outward manifestations of values by which we *perceive the organizational culture*. Whatever the perception, positive or negative, we have recognized (or judged there to be) a certain uniformity of attitudes and patterns of behavior.

What we are trying to get across is the fact that, with or without intent, the dean of the college, the director of the laboratory, or the top manager of an organizational unit sets a tone which reflects the values held important to the people who make up this organization. The success of a person in such a position is based upon a pervasive awareness of the effect of her or his behavior upon the perceived cultural values of the organization.

Organizational culture, in the positive sense, is a most valuable and necessary asset. When it is lacking, it is generally perceived as a negative asset. It suggests that people with negatively perceived values have set the tone.

13.7   Relevance to Creative Effort

Why is organizational culture particularly relevant to the management of creative effort?

In maintaining the flexibility required to facilitate creative effort in the world of high technology, rules and regulations can get in the way. The challenge of new initiatives, new opportunities, and new organizational structures demands rapid and spontaneous response. The often intensely competitive nature of such effort places considerable pressure upon the participants. In the absence of specific guidelines to deal with a given situation, a clear perception of an organizational culture is essential.

This organizational culture is to be perceived much as the benefits of a good upbringing: A set of values and behavior acquired by emulation, which serve us well at home and when we are away from home. As in the family, such values evolve in an organization as a result of shared experience, not through preaching or admonition. It may be necessary to create or change the rules designed to preserve these values in the face of changing conditions. The basic values must have staying power.

A new member of an organization should be prepared to accept these values. Without an explicit set of such values, the individual is left to drift in turbulent waters without navigational aids. At best, creativity is inhibited by making people choose a safe course to avoid real or imagined obstacles. At worst, people find themselves facing problems which may put their personal integrity at risk.

Organizations cloaked with a mantle of "creativity" (maybe not the sort you would approve of) continue to operate in our society. I have encountered a "successful" Hi-Tech company, which, for more than 20 years, has been able to attract new investment capital without having produced a product or a profit. The company employs more than 200 people, and its stock is traded publicly. This company has attracted key employees over the years through excessive salaries (more than twice current market rate). It operates without a formal management structure.

All decisions are made by the CEO, the major stockholder. The implicit strategy is to allow perpetuation of the absolute control of the organization by the CEO. He has been singularly successful in attracting investment capital, which to date has sustained the organization. Without this individual, this organization would collapse. An explicit set of organizational principles would defeat the very essence of this operation: it would have to explicate the operational basis, i.e., self-deception on the part of the CEO which is the basis for his success in attracting investors.

This is an example of one of the extremes which continue to support some corporate existence. It has taken from 1 to 3 years for mature and experienced employees to recognize the significance of this implicit organizational culture before they left the firm. Young scientists and engineers lacking relevant experience may not recognize the implications of such an organizational culture at all. They may well assume that this is the way of the world. The lack of a formal management structure and the concentration of decision-making power in a single individual can serve to hide a multitude of sins. Caveat emptor.

13.8 Summary

An attempt has been made to establish the need and basis for cultural values in an organization. Values such as trust, integrity, and respect for others are highlighted through illustrative examples. Among the most relevant is dealing with intellectual property. The unique role of cultural values in creative enterprise has been pointed out. Both sides of the concept of bureaucracy are presented and the relevance to creative effort is emphasized.

14. The Impact of Globalization and Technology

Globalization and technology have been the principal agents of change for the past 20 years. This has not only affected our daily lives, but has also brought new challenges to the management of Science and Hi-Tech. What are the implications of that change?

14.1 Globalization

"Globalization, outsourcing, and information and communications technologies (ICT) have combined over the past 10 years to basically level the playing field for the R&D community. Technology organizations can now outsource the development of new technology-based products to various countries that have many of the technology resources once limited to advanced economies—the USA, Japan, and Europe. Emerging economies that include China, India, Korea, Brazil, and Eastern Europe are now able to compete with the former technology triumvirate for development of the most sophisticated and technologically complex new products."

From a Batelle report by M. Grueber, R&D Magazine, Dec. 2010

http://www.battelle.org/aboutus/rd/2011.pdf[1]

To fully appreciate the *magnitude of this impact*, here are some numbers from the same Batelle report:

2009/2011 Global R&D Funding: $1,107/*1,192 Billion*		
Of this total, the respective share (percent of total):		
Americas 39.1/38.4	Europe 24.1/23.2	Japan 12.6/12.1
China 11.2/12.9	India 2.5/3.0	Other 10.3/10.3

(Note an average YOY change $+7.2\%$ for China and India and -1.6% for all others.)

---

[1] See also:

Globalization and Offshoring of Software:

A Report of the ACM Job Migration Task Force.

William Aspray, Frank Mayadas, Moshe Y. Vardi, Editors,

Association of Computing Machinery, 2006

R. Kay, *Managing Creativity in Science and Hi-Tech*,
DOI 10.1007/978-3-642-24635-7_14, © Springer-Verlag Berlin Heidelberg 2012

Not only that R&D in China and India has reached its present level in just one generation, but the associated recent rise in patent activity also suggests that this R&D effort has gone from the catching-up, to the innovative phase (Wright 2007). Given the size of China and India's population, the implications for the future are clear.

What is more, many major corporations with a large Hi-Tech component have established or expanded their international research presence (Bouderi 2000).

NEC of Japan has laboratories in the USA and Germany; Siemens of Germany has had a laboratory in Princeton, New Jersey, since the 1980s; Xerox has laboratories in France, England, and Canada. IBM has added laboratories in China, Israel, and India to those in Japan and Switzerland. Novartis has 20 R&D laboratories dispersed in Europe, Asia, and the USA. Microsoft has laboratories in England and China. General Motors has laboratories in Germany, Israel, India, and China.

Huawei Technologies, founded in 1988, has laboratories in the USA, India, Russia, Australia, Sweden, and Mexico.

For a comprehensive analysis, see Globalizing Industrial R&D,

U.S. Office of Technology Policy, 1999

http://www.dudebin.com/library/Dalton.pdf

14.1.1 Management Implications

Successful transnational corporations have learned to recognize the unique needs of customers in various parts of the world. They have found ways to make use of local suppliers and service providers. Outsourcing of global services has added another level of complexity.

Managing R&D on a global scale is a new challenge. To balance the multiple interests inherent in a European Union project made up of different kind and size of organizations from various countries is one example of the scope of issues to be faced.

Buderi's account of the evolution of R&D makes clear that there are many ways to achieve success,—or to encounter failure, for that matter (Buderi 2000). The evolution he describes has materialized over a period of more than a 100 years. Asia's role in global R&D has emerged in one generation—with the potential of overwhelming magnitude.

(To quote the preface to this edition:)

The new leaders in Science and Hi-Tech (in Asia) have an advantage which is hard to overestimate: In the course of their education or early experience in the USA or Europe, they have come to accept that there is "another way of doing things," the essence of creativity. They have had the experience of adapting to another culture—and selecting that which promises to serve their *future* needs. They are less burdened by the notion born from sustained success:

"We have always done it this way."

Global competition in the world of R&D will make new demands upon management in what are now the leading nations. Even seasoned scientists and engineers with experience in the management of large projects most likely have not been exposed to problems of international scope.

The establishment of R&D operations outside of the home country has created opportunities to remedy this situation. Promising R&D managers can be offered career opportunities abroad. ("Abroad" is here defined relative to the country in which first hired.)

The significant number of foreign-born top level R&D managers in Silicon Valley suggests that the search for excellence has become global.

The problems encountered by Google's China operation (Levy 2011) underline the contention that globalization can defy even a forward-looking and successful Hi-Tech management team.

14.1.2 National Differences: Changing Perceptions

Some earlier beliefs are being replaced:

It was long assumed in the West that capitalism and innovation could flourish only in a democratic society, i.e., a society perceived to be conducive to change.

In 1989, octogenarians made up a substantial portion of the delegates in the People's Hall in Beijing while their great-grand-children's protest was being subdued outside, on Tiananmen Square.

One generation later, octogenarians had disappeared from the scene and China had become the second largest economy in the world. At the same time, Deng Xiaoping, credited with the "Socialist Market Economy," has raised the standard of living of hundreds of millions of Chinese.

This represents *unprecedented* change in the history of the world, on the part of a country with a culture going back more than 4,000 years. China may have a way to go to catch up with the West,—there can be no doubt that it will.

Anyone who believes that Microsoft's Bill Gates and Mark Zuckerberg of Facebook are unique to American culture should learn about Alibaba's Jack Ma, a young English teacher from Hangzhou who brought e-commerce to millions of Chinese, connecting China to the world and beating eBay, a $9 billion company, at its own game.

Much credence was once given to the idea that Germany's relative low standing in the world of Hi-Tech was due to the rigid traditions of its academic and banking institutions.

In fact, industrial research had its beginnings in the German chemical industry. Already early in the twentieth century, innovative German chemists were among the highest paid members of their companies, by way of profit sharing plans, based on individual patents.

At the end of World War II, German patents were expropriated by the victorious Allies. Furthermore, war and emigration had led to a severe depletion of the ranks of qualified faculty in science and engineering. And investment in a war-torn infrastructure had to take precedence over the pursuit of new technologies.

The USA and Britain, on the other hand, had a surplus of well-trained people (including those who had left Germany). They were available to staff faculties of emerging fields, such as computer and material science. These new fields attracted some of the most enterprising mathematicians, aeronautical engineers, physicists, and chemists.

The USA found itself in the unique position of being able to provide a war-torn world with both the capital and the capital goods to rebuild. The ensuing 50 years of US-led economic growth made possible unprecedented advances in Science and Hi-Tech.

Two generations later, there is evidence that efforts to catch up are making headway in Germany (Lehrer 2000). Data reflecting Venture Capital activity for the period 2000–2008 place Germany ninth among the top ten countries, in terms of successful start-ups (Brander et al. 2010). Scheer, a well-respected German Hi-Tech manager, goes so far as to say that the relative success of South Korea and Malaysia vis-à-vis Germany is due to the shortage of Hi-Tech managers in Germany (Scheer 2010, p 13).

There is another way of looking at national differences which account for a more or less favorable climate for creative endeavor and innovation.

Educational policy and government regulation of entrepreneurial activity, as well as tax and immigration policy, have an impact upon the development of Science and Hi-Tech.[2]

Since the principal source of funding of higher education comes from national and state governments, national priorities clearly influence allocation. For example, some European countries have put equal access to education ahead of building elite institutions of higher learning.

In the USA, private universities such as MIT, Stanford, Harvard, CMU, and Cal Tech have played a major role in advancing science and technology along with the major public universities.

National tax policy plays a large part in this. In the USA, universities get considerable financial support from tax-deductible gifts from alumni and corporations. In addition, there is corporate and federal government support for explicitly defined research. Tax-favored foundations provide alternatives to peer-reviewed grants, one of the few ways to embark upon game changing innovation (see Chap. 12, Financing Creativity).

Traditionally, the USA has had a more open immigration policy than most European or Asian countries. After all, the USA is a relatively young nation of immigrants.[3]

---

[2]The above-quoted study by Mark Lehrer is an extensive, carefully researched, and evenhanded analysis of an effort to influence and direct Hi-Tech activity in Germany over a 10-year period.

[3]Israel is a small immigrant country with a unique history, which lends weight to the notion that diversity correlates with successful Hi-Tech effort.

Europeans (be they English, French, Italian, or German), or Chinese, Indians, or Russians are likely to find communities of fellow country-men and -women, a significant factor in choosing to settle in the USA. In these communities, they find not only schools where their children can get a bilingual education, but also shops that cater to their dietary preferences as well as churches that provide for their spiritual needs.

In turn, these communities, often associated with university clusters, have attracted Hi-Tech industries and thus offer jobs and career opportunities as well. There have been efforts to emulate Silicon Valley in other parts of the USA, in Europe, and most recently in Russia and China.

The experience of the San Francisco Bay Area, of which Silicon Valley is a part, suggests that it takes more than a generation for such communities to evolve, even in a climate of supportive government policies. While Silicon Valley looks to be on solid footing, there is no guarantee that such communities can be sustained. Star athletes aspire to the Olympics, irrespective of its geographic location: which probably is one reason that it has been so difficult to replicate Silicon Valley (Saxenian 2002).

In some countries, the attitude among academics has restrained serious involvement with Hi-Tech endeavors springing from their respective fields of science. The custom of delaying the granting of advanced degrees to the most talented "assistants" has deferred their participation in "startup" companies. Where these attitudes have been overcome, the results have been remarkable.

The growing popularity of English language programs in the universities of non-English-speaking countries designed to attract foreign students will further ameliorate disparities.

The effectiveness of the top–down approach to promoting Hi-Tech enterprise as exemplified by Russia's Skolkovo initiative is yet to be demonstrated (Kramer 2010). The motivation to develop human resources is greater when there are no natural resources to exploit (Zaytsev 2011). Yet, there is reason to believe that the Russian software industry shows promise (Bardhan and Kroll 2006).

Comparison of national differences, on a less subjective basis, has been attempted.

Probably the most acclaimed multinational study of work-related values by Hofstede (1980) makes a convincing case for culture-based values which suggest the need for culture-adapted management styles. (This study has been extended since to a large part of the world.)

The Science and Hi-Tech environment does not bear this out, at least in my experience. The mobility of Science and Hi-Tech personnel among the industrialized nations has created—an almost universally adapted—modus operandi which makes it difficult to identify the geographic location of an R&D laboratory with anything but the food in the cafeteria. (If not universally adapted, at least tending in that direction). Where 30–40 years ago the need for a native laboratory director was held to be imperative, today her or his professional stature is the determining qualification. (This does not invalidate the dictum that top management of a foreign subsidiary be nationals of the subsidiary location rather than nationals of the headquarter country.)

Very likely, the most glaring exception to the view that divergent cultural values have little influence upon the world of Hi-Tech is the experience of Google in China (Levy 2011).

A survey of the venture capital industry worldwide cites unfavorable tax policies and an unstable regulatory environment as its most pervasive impediments; the lack of entrepreneurial talent is perceived to be the greatest in Germany, Canada, China, UK, France, and India, and least in the USA and Brazil (Deloitte and NVCA 2010).

This view is reflected in a detailed academic study of the actual number of VC funded start-ups in the ten most active countries (Brander et al. 2010).

14.2   Impact of Technology

Having taken note of the impact of globalization, it is imperative to recognize the impact of technology not only upon every aspect of contemporary life but especially upon creative endeavor and the practice of management itself.

E-mail and the Internet have revolutionized the dissemination of information which is the lifeblood of creative activity. Instant communication within organizations and across continents has made possible collaboration on an unprecedented scale. Just the idea that multiple authors can collaborate on the creation of a publication without the impediments of time and distance has changed the way information turns into knowledge. Creative effort has been enhanced through the ensuing ability to avoid duplication and the early recognition of failed approaches. While it has made many disciplines more competitive and life a bit more hectic, the benefit to society has been immense.

To gain perspective—China: 1987.

- Imagine yourself at a provincial airport. You are waiting for the arrival of a scheduled, overdue flight. No one is able to give you any information.—The only communication link is the airplane's radio which has a range of 50 miles.

- The river barge is still the principal mode of bulk freight traffic in China. You notice empty barges in both directions. - There is no timely way of getting data about waiting cargo to the returning barges.

- Founding of Huawei Technologies in Shenzhen.

By 2010, 24 years later, Huawei is the world's second largest supplier of Telecom equipment.

Nothing has contributed to China's rate of progress more than the availability of wireless broadband communication.

Today, management can react to a problem with an instant message to possible responders scattered across the globe; but this idea is only beginning to be developed to its full potential. *Enterprise 2* will surely come to encompass every aspect of our working life much as social networks have assumed a significant role in our social interactions (NY Times, 6/27/2011,B3).

If one accepts the premise (advanced in Chap. 8) that you should hire the brightest you can find, it behooves you to find ways to get their input to the widest range of issues in the most efficient and timely manner. The technology to do this is now available.

Google has shown one of the most dramatic examples of game-changing technology: Subject the most massive accumulation of machine-readable data, the World Wide Web, to statistic analysis and you change the way the world goes about its business. In the course of a few years, it has brought radical change to the advertising business and has turned itself into one of the most profitable large companies, ever (Levy 2011).

The long pursued goal of machine language understanding has made decisive strides; this implies that we will soon be able to ask questions with the expectation of meaningful answers, rather than possibly relevant reference material.

The promise of new technology in the pursuit of solutions to totally unrelated problems has been one of the hallmarks of creativity.

Advances in planar technology, the foundation of semiconductor devices, have also been the basis for major advances in printing and computer storage products: the ink jet nozzles and modern recording heads have allowed these products to keep pace with other computer system components. Planar technology is also making inroads in biotechnology by offering potential solutions based on related nanotechnology.

The increasing role of *open source* software has contributed to the worldwide availability of Hi-Tech products and services.

Even a routine task of scientists and engineers, the challenge of a well-formatted visual presentation, has been largely met by aids such as *PowerPoint* (see Chap. 10).

Video conferencing offers a valuable tool in facilitating the management of geographically dispersed projects.

The Internet Portal *AcademiaNet* (http://www.academia-net.de) makes it easy to identify top female researchers in Germany; it aims to simplify the process of finding the right candidates to fill science policy commissions—where women are often seriously underrepresented.

While the creative contributions of science and technology have caused their share of problems, in balance the benefits have outweighed negative consequences.

The world is feeding more people better than it did 100 years ago. For the great majority in the industrialized nations, the quality of life has improved in terms of health, longevity, education, and creature comforts, reflected in a rise in real per capita GDP of nearly 800% (Baumol 2008).

> The potential for global cooperation toward peace and greater prosperity has been enhanced by advances in science and technology. Whether we have the wisdom to make use of these advances for the benefit of all life on earth remains to be seen.

The near collapse of the world's financial system during the first decade of the twenty-first century, produced by reliance upon management by "financial engineering," was an ominous warning. The reluctance to accept a predictive macroeconomic theory implies a mindset on the part of economists which has not served us well (Danielmeyer 1997).

Scientists and engineers must assume greater responsibility for the welfare of their organizations, their countries, and the health and resources of this planet which we all must share.

There is growing awareness of the need for the management of creativity in Science and Hi-Tech in many parts of the world (Humboldt-Kosmos 2009).

National or regional differences have led to different views as to the specific manner in which this need can or should be satisfied. There is little doubt that once recognized, multiple paths will be found to answer this need (Thomas 2005).

The role played by some of the most esteemed US scientists and engineers who spend a few years in government service is a striking example of their potential impact. They are recruited from research universities and leading industrial research laboratories and provide the government with the kind of experience and insight not readily found among career civil servants. They are involved in developing *National Policy for science and technology* and take responsibility for prioritizing and allocating a large part of the total R&D expenditures. Their effectiveness is based upon their status within the scientific community, facilitated by personal contacts. The organizations, which furlough a highly valued member for such public service, benefit from the widened perspective and new contacts gained by the returning individual.

This practice of temporary assignments which has facilitated technological leadership has *not* been widely emulated outside the USA.

14.3   The Future

An increasing number of the world's problems are within the scope of responsibilities which fall upon managers of Science and Hi-Tech. It may be some time before the world recognizes the long-term futility of solving these problems within the confines of national self-interest.

The possibility of major pandemics, nuclear accidents, the consequences of the growing demands for energy, the effects of major natural disasters upon high-density population centers, and disruption of increasingly more complex communication systems upon which we are ever more dependent are examples of the challenges ahead.

In terms of time and available resources, the boundary conditions imposed upon potential solutions suggest the need for prioritization and collaboration on an international scale.

In some parts of the world, a number of nations have joined forces to sponsor efforts to meet the challenge. The Nordic Alliance of countries in the North Atlantic and Baltic Sea is a notable example.

In the European Union, the FET Flagship and ERC initiatives are other efforts to overcome the constraint of limited resources on the part of some member states. Insofar as the objectives are similar to programs elsewhere, it will be interesting to see how well such collaboration will fare competitively.

My own experience with such efforts, involving many diverse organizational entities, suggests effective leadership and managerial skill to be the critical path to success.

The spontaneous uprising by young people in the Near East in early 2011 has demonstrated that our interconnected world is subject to unprecedented change, seemingly beyond our ability to anticipate or accommodate.

Many of the ideas which have served the industrialized world so well were conceived several hundred years ago. The way we think about some of the related issues is hopelessly outdated in terms of future needs, such as environment, healthcare, education, financial system, security, and transportation.

Insofar as science and technology are to play a significant role, creative scientists and engineers must rise to the challenge. These challenges to scientists and engineers at all levels of management and the ensuing opportunities were never greater.

14.4 Summary

Having identified the magnitude of the impact of globalization, we have drawn attention to the global redistribution of future potential and to the changing perception of national differences.

By looking at the recent impact of technology upon the world and the implications for Science and Hi-Tech itself, a case is made for the challenges and opportunities confronting tomorrow's managers of Science and Hi-Tech.

15. What the Behavioral Sciences Have to Offer

The objective of this chapter is to motivate further study of this subject. For a person educated in engineering or the natural sciences, this can serve as a rewarding introduction into the behavioral sciences because of its immediate relevance to the work environment.

Remarkably, the interest of behavioral scientists has only turned to the subject of people at work and the behavior of groups in modern times. Not until the late 1950s did psychologists and sociologists venture into the factory and office to find out what makes organizations tick. Of those who had a major impact, the work of McGregor, Maslow, Herzberg, and Likert seems most relevant.[1]

15.1 Douglas Mcgregor: Theory X and Theory Y

McGregor, one of the most influential social scientists concerned with management, was among the first to challenge the widely held view that people dislike work, need to be coerced to work, and prefer to be directed in return for security (McGregor 1960). McGregor called this earlier view, Theory X, and categorized it as a self-fulfilling theory; a theory which describes what happens if you accept Theory X.

With Theory Y, McGregor proposed that:

1. Work is natural
2. External control through the threat of punishment is not the only way to motivate people
3. Commitment is a function of the rewards associated with individual achievement

---

[1] Some of the material in this section is based upon the Conference Board publication "Behavioral Science; Concepts and Management Application," by Harold M.F. Rush (1970), an excellent introduction to the subject.

4. Most people are capable of a higher degree of creativity in solving organizational problems, and
5. Their intellectual potential is generally underutilized

Central to McGregor's work is the importance of employee commitment and its direct relationship to motivation. People are capable of creativity, self-control, and integrating their goals with those of the organization, but it does not follow that they will become committed and involved. One important determining variable is the degree of emotional maturity of the individual and at what level he or she is motivated.

While many of his followers accepted McGregor's theories as a set of principles, he himself saw them merely as a set of premises to be tested in a given situation for their applicability. He recognized that *personal motivation* is a complex and situational matter, a point which was brought home to me in a very dramatic fashion.

I was on a jury which had just heard a young man accused of drunken driving admit that he had been drunk when apprehended. His defense was that the arresting officer "had it in for him." A clear case of guilty. The jury entered deliberations at about 5 pm and as is the custom, each juror in turn gave a proposed verdict and the basis for her or his decision. All but one of the jurors found the accused guilty. While waiting my turn, I tried to analyze the possible motivation for the lone dissenting vote. I wanted to counter her argument in a way which would bring the deliberations to an early conclusion, so we could all get home for dinner. She argued a case where a police officer had been looking for an excuse to arrest someone in her family. I was not able to change her mind in spite of the force and logic of my argument.

She changed her vote only after two successive polls and when all the other jurors had turned against her.

As the foreman prepared to take the jury's verdict back into the courtroom, he casually remarked that if this had taken 10 min longer, we all would have had dinner at the expense of the court. To which the lone dissenter countered: "It isn't that I didn't try."

My own motivation, to reach a verdict as quickly as possible, had prevented me from considering an alternative motive with which I could not identify. She evidently did not have someone at home waiting with a dinner prepared for her.

Returning to McGregor's theory X and theory Y, these theories are based on different perceptions of human behavior, the former more negative, and the latter more optimistic. Undoubtedly the increasing prosperity of the industrial society over the past 100 years has contributed to the acceptability of the more optimistic view.

To appreciate the significance of McGregor's contribution it must be borne in mind that these were new and radical postulates in their time. While they are generally accepted in many parts of the world today, there are still places even in the industrialized world, where these concepts have not replaced culturally rooted concepts of people and work.

15.2 Abraham Maslow: Hierarchy of Needs

Maslow's convictions about a society populated by superior human beings has had a profound influence upon the developments of the late fifties and early sixties. His use of the term "self-actualization," as a rarely obtained state of perfect human achievement, gave impetus to the notion that there are unfulfilled human needs which provide motivation of the highest order.

No one in the field has had a more significant and lasting influence upon modern management than Maslow. His ideas are particularly relevant to creative enterprise (Maslow 1954). He is best known for his Hierarchy of Needs, a concept which has found wide acceptance

> - Need for self-actualization
> - Need for esteem
> - Need for belonging-ness and love
> - Safety needs
> - Physiological needs

The lowest level of needs, termed physiological, refers to food, warmth, shelter, elimination, water, sleep, sexual fulfillment, and other bodily needs.

Safety needs include actual physical safety, as well as a feeling of being safe from injury, both physical and emotional.

The need for belonging-ness and love represents the first social need. It is the need to feel part of a group or the need to belong to and with someone else. It implies the need to both give and receive love.

The need for esteem is based on the belief that a person has a basic need for self-respect and the esteem of others: first, there is a need for feeling a personal worth, adequacy, and competence; second, there is the need for respect, admiration, recognition, and status in the eyes of others. (The discussion of nonmonetary compensation in Sect. 8.3.5, addressed itself to this need.)

The need for self-actualization is more difficult to describe. Self-actualization is the process whereby one realizes the real self and works toward the expression of the self by becoming what one is capable of becoming, i.e., by achieving one's potential.

The focus of this book, Managing Creativity, attaches particular significance to this last need: Creative people's desire to choose, influence, and control the content of their work; it is for them the most important aspect of self-actualization.

According to Maslow these needs occur in a hierarchical sequence throughout a person's development and maturation. The hierarchy underscores a fundamental point: Until one need is fulfilled, a person's behavior is not motivated by the next, higher-level, need. By the same token, once a need is satisfied, it no longer motivates.

Since the lower-level needs are the most urgent ones, they must continually be satisfied in order for a person to be motivated toward higher level needs. But even when the lower level needs are satisfied, if they become threatened, they again become a source of motivation. A man who is safe may risk his safety, even his life, if he becomes hungry enough. Or a person whose prime motivation has been esteem may drop down a level to seek belonging-ness, if the latter is threatened. Loss of a place in the starting lineup is generally preferable to not making the team.

For Maslow, self-actualization shows evidence of emotional maturity. And by its very essence, self-actualization is a *self-perpetuating, ongoing, and never finished process*. Its most important characteristic is that unlike other sources of motivation, which extinguish themselves after the needs are fulfilled, self-actualization continues to motivate people to ever higher levels of performance.

To quote Rush (1969a): For Maslow, self-actualizing people are motivated to involve themselves in creative expression that they themselves find gratifying.

This notion is basic to our concept of managing creativity: To protect and nourish an environment which supports creative enterprise. In some situations, creative people will forgo the satisfaction of lower level needs, in order to maintain control of their creative effort (Sect. 3.6 Individual Creative Effort).

15.3 Frederick Herzberg: Motivation-Hygiene Theory

Herzberg built his motivation-hygiene theory on the results of a study which questioned the job attitudes of 200 accountants and engineers (Herzberg 1966). He asked: "Can you describe, in detail, when you felt exceptionally good about your job?" and "Can you describe, in detail, when you felt exceptionally bad about your job?"

Analysis of the responses showed that positive responses were generally associated with job content, labeled content factors, or *satisfiers*; experiences associated with bad feelings were associated with context factors or *dissatisfiers*.

He found *the job content factors*, reflecting satisfying experiences, to be:

- Achievement
- Recognition
- Work itself
- Responsibility
- Advancement
- Growth

He categorized *the context or environmental factors* causing dissatisfaction to include:

- Company policy and administration
- Supervision
- Working conditions
- Interpersonal relations
- Salary
- Status

- Job security
- Personal life.

The detailed definition of these factors provides considerable insight (Rush 1969b). For example:

Responsibility: Both responsibility and authority in relation to the job are included here. Specifically, "responsibility" refers to the employee's control over his own job, or to his/her being given responsibility for the work of others. This factor is different from the consideration of whether or not there is a gap between a person's authority and the authority needed to carry out the relevant job responsibilities. When this gap was reported, Herzberg classified it under "company policy and administration," on the assumption that the discrepancy was evidence of poor management.

Job security: The criterion for this category is objective signs of the presence or absence of job security, not feelings of security. Responses under job security include tenure and company stability or instability.

Herzberg also called the dissatisfiers *hygiene* factors based on the medical definition of hygiene as "preventive and environmental." The satisfiers, which are all related to the job itself, were called motivators, since other findings in the studies suggest that they are effective in motivating employees to greater performance and productivity. The original studies of 1959 have since been replicated numerous times, including cross-cultural studies in other countries. All led to findings which closely parallel those of the original study.

These early pioneers forced revision of generally held beliefs about the nature of motivation. They also introduced some elements of quantitative analysis to the research process.

15.4 Rensis Likert: Four Management Styles

While Likert made a significant contribution conceptualizing four management styles (Likert 1967), he is best known for his introduction of an attitude-measuring device which has gained popularity as the "Likert-type scale." For illustration, here are some typical questions on a Likert-type scale:

1. The medical benefits of this company are not liberal enough.

   (a) strongly agree
   (b) agree
   (c) no opinion
   (d) disagree
   (e) strongly disagree

2. My manager is a forceful leader.

   (a) strongly agree
   (b) agree
   (c) uncertain

(d) disagree
(e) strongly disagree

Each of the responses is given a numerical weight, the most positive reaction usually weighted as "5."

Likert proposed to apply such quantitative measures in evaluating an organization's management style. He conceptualized four management styles:

System 1. Exploitive-Authoritative
System 2. Benevolent-Authoritative
System 3. Consultative
System 4. Participative-Group

According to Likert these four styles and their accompanying organizational climates all exist in everyday practice, and any functioning organization can be characterized by one of these. He offers a body of research to support his statement that the *System 4.* company is the most creative, successful, and effective.

To determine the management style of a given organization, Likert developed a set of 22 questions which form the basis for his "Comparative Analysis of Organizational and Performance Characteristics of Different Management Systems." To illustrate, we have chosen a few of the questions:

- To what extent do superiors have confidence and trust in subordinates?
- What is the amount of responsibility felt by each member of the organization for achieving the organization's goal?
- What is the direction of information flow?
- How well does the superior know and understand problems faced by subordinates?
- To what extent is technical and professional knowledge used in decision-making?

To each question, Likert offers a set of four possible answers which are associated with the four management styles. In selecting the answer which best reflects the organization to be evaluated, a surprisingly consistent and revealing profile emerges.

15.5 Summary

This very brief treatment of the subject can only serve as an appetizer to some rich fare. As must be evident, the ideas identified with these pioneers have influenced my perception of what the management of creativity is all about. The occasional lectures by some of these individuals, which I have been able to attend, have helped me overcome my prior skepticism toward the behavioral sciences. This skepticism, based largely upon lack of familiarity and imagined lack of relevance, is probably shared by many engineers and by scientists from fields other than the social sciences.

The seminal contributions of these four distinguished behavioral scientists have been chosen for their relevance and as an introduction to the subject, intended to stimulate further interest.

McGregor was one of the first to recognize that personal commitment and motivation are important, complex, and situationally conditioned factors in people's attitude toward work. Thereby he changed the views which prevailed as little as 50 years ago.

Maslow contributed the concept of a "Hierarchy of Needs," the highest need being that of "Self-actualization," which is particularly relevant to the pursuit of creative effort.

Herzberg refined the concept of needs. He draws a distinction between those needs which primarily satisfy when met and those which primarily dissatisfy when not met. The former tend to be associated with the content of the work, and the latter with the context in which the work takes place.

To *Likert* the field owes the introduction of quantitative methods and their application to the recognition and categorization of organizational management styles.

16. Management Training Opportunities for Engineers and Scientists

This overview of relevant management training programs is intended to provide the reader with some perspective on what to look for. Availability will vary significantly with geographic location (an issue not addressed). E-Learning opportunities have helped overcome this problem.

16.1 Short Programs

Many large organizations provide training opportunities that address management issues which are unique and specific to an organization or department function. Most often, they are tailored specifically to the administrative, sales, or manufacturing functions.

The business or management schools or departments of numerous universities and colleges provide both function-specific and "General Management" courses for people who have entered the work force. The duration is from one to several weeks. Some of the better known business schools, including Harvard, MIT, Wharton, and Stanford, also offer 1-semester or 1-year courses, primarily for experienced managers. Leading business and management schools the world over offer such curricula patterned after these early leaders.

There is a large variety of courses offered by private companies which provide management training. They frequently have university faculty among their staff. Professional societies occasionally sponsor or endorse such courses.

One-day courses on anything but a very specific aspect of management should be considered with caution.

General management training, i.e., not function specific, usually takes one of two approaches, and occasionally a mixture of the two:

- The most widely available is the classroom setting where the emphasis is upon presentation of material by the lecturer.

R. Kay, *Managing Creativity in Science and Hi-Tech*,
DOI 10.1007/978-3-642-24635-7_16, © Springer-Verlag Berlin Heidelberg 2012

- The alternative approach involves active participation by the members of the group. This approach is sometimes labeled as sensitivity or awareness training or the assessment method.

16.2   Basis for Selection

Having been exposed to a large variety of management training programs as a working scientist, engineer, and manager, I have always gained something from such programs. In evaluating their effectiveness from the viewpoint of the engineer or scientist contemplating management or having recently assumed management responsibility, some general issues can be identified:

1. Many or most such programs are aimed at people who represent the median of intellectual discipline as measured by overall academic achievement. People who come from the most demanding university graduate schools generally find such programs slow moving, and consider them lacking in concrete content in relation to the time invested.

On the other hand, a significant value of any program that draws people from diverse backgrounds lies in the opportunity to meet people from different disciplines (from outside or within the organization) and to learn to appreciate their approach and viewpoint.

2. The quality of training programs involving group participation is likely to be inconsistent. Such programs are particularly dependent upon the individual instructor: Not only the instructor's knowledge and delivery, but also his/her ability to manage a particular group. Hence, a priori evaluation of such programs is difficult.

The lecture format is more often offered by universities, where faculty membership provides some basis for continuity and thus for judging excellence. This format allows for relatively large classes.

> The best-informed decision regarding participation is usually based on the opinions of colleagues and/or managers who have participated in a particular program.

16.3   Career Development Workshops

The most successful approach, in my experience, is the 1 week workshop designed to give highly motivated engineers and scientists an opportunity for self-assessment of their interpersonal and administrative skills.

The measure of success which I have applied to this program is its voluntary self-perpetuation over more than 15 years in an environment in which other voluntary programs aimed at management training have fallen by the wayside.

It is an expensive program in terms of people-time. It involves six "participants," usually nonmanagers likely to be receiving management assignments in the near future, or inexperienced managers. It also involves six "observers," generally managers with several years of management experience and two or three professional personnel (HR) people. The experienced managers act in the role of observers and are responsible for the generation of individual reports on each of the participants. Experience has shown that the "observers" gain as much from the program as do the participants. The professional personnel people are responsible for the conduct of the program.

Over a number of years, it is possible to expose the most promising individuals with a transformative experience with a marked effect upon the organization.

A unique (and to some, controversial) aspect of this program has been the notion that the reports and observations about participants are NOT made available to the management of the organization, i.e., are for the benefit of the participants only. This is based upon the strong conviction that an assessment program, designed to serve management in its evaluation of people, creates an attitude on the part of the participants which distorts their behavior. They are likely to act or react in a manner designed to reflect "favorably" upon themselves.

The program is very intensive, challenges the participants continually, and all of the material used is very much *tailored to the environment and interests of the participants*. For example, there is an In-Basket Exercise in which each participant is confronted with material typically found in the laboratory manager's in-basket. This exercise, incidentally, makes the scientist or engineer aware (usually for the first time) of the problems faced by the director of the laboratory.

The Management Selection Task is based on material directly relevant to the environment from which the participants are drawn. The material often relates to an emotionally charged issue of current concern. The exercises are designed to challenge the participants individually and as members of a team. Some issues are presented first to the participants individually. Subsequently, all participants are asked to reach a consensus position, followed by a review of the group's approach to reaching consensus.

It is interesting to observe that in the course of such a series of related tasks, the problem defined at the outset undergoes significant change. You may start out with an ostensible technical problem which morphs into one of conflict of interest.

The matter of personal interaction among the participants is very much brought into focus. This is not easy to accomplish in a nonthreatening way, hence the need for experienced professionals in the conduct of such a program. The observers provide continuous feedback to the participants. The final report, prepared for each participant, is based on the collective observations of the staff and represents the consensus of all observers.

The relative success and enthusiastic acceptance of this program among advanced-degree professionals suggest that most management training programs aimed at a broader class of individuals fail to engage the people who constitute this very important target group in the Hi-Tech environment. The results of the Career Development Workshop referred to above have been evaluated by Gary Hart (1977).

16.4   Full-Time Academic Training Programs

By virtue of proximity to and long-term involvement with Hi-Tech organizations, Stanford and MIT have become the most renowned schools of management to address the unique needs of Science and Hi-Tech. Many schools the world over have followed their lead.

The MIT Sloan Fellows Program in Innovation and Global Leadership is representative of the most highly rated offerings. http://mitsloan.mit.edu/fellows/overview.php

It has evolved from its precursor, the MIT Management of Technology Program which was initiated at the MIT Sloan School of Management and School of Engineering in 1981.

This program is aimed at employer sponsorship of students; the tuition fee is $119,000 (June 2011). Its success is evidence of the fact that industry and government recognize the need for management training of technical personnel. *One of the principal benefits is the year-long association with a highly selected group of people with similar experience and interests.* This is a 12-month, full-time program leading to an MS in Management of Technology.

Admission requirements:

- An undergraduate technical degree
- A minimum of 5–10 years' experience in industry or government in areas related to technology
- A year's work in calculus and economics.

Courses offered in summer, fall, and spring:

- Applied Micro- and Macroeconomics
- Statistics for Management
- Financial and Management Accounting
- Engineering and Systems Analysis
- Managing Professionals
- Seminar and Workshop in Computer Systems
- The R&D Process: Communication and Problem Solving
- Marketing Management
- Technology Strategy
- Strategic Management
- Manufacturing/Technology Interface
- Coordinated Strategies for Managing Research, Development, and Engineering
- Government and the Management of Technology
- Thesis Research

Some of these courses reflect the research interests of the faculty, drawn from the management and engineering schools.

16.4.1 General MBA Programs

Universities with strong science and engineering programs often have MBA programs that involve faculty from the school of engineering, which make their MBA programs more relevant to science and engineering graduates. Some also offer a combined degree in technical management.

Some universities in every major metropolitan area offer evening MBA programs. These usually involve attendance over 2–4 years.

16.4.2 E-Learning Programs

A number of large Hi-Tech organizations have encouraged motivated engineers and scientists to avail themselves of the so-called Mini-MBA programs offered by major universities online. Recommendation by a trusted colleague is the best way to assure a good match.

Some organizations have developed e-learning programs which address specific aspects of management training, unique to a particular function. Their growing popularity implies that they meet a real need and have proven themselves to be effective. Their advantage lies in the fact that you can sample the offering, in situ, and proceed on the basis of your own assessment.

16.5 Task Force Participation

Large and medium size organizations have a unique asset in the form of experienced managers of diverse backgrounds as well as experts from various disciplines. Together they constitute a pool of potential consultants with experience relevant to the organization, difficult to match by outside consultants. The most task-relevant members of this pool can be assigned temporarily, to make up a task force charged with addressing a specific problem.

Participation as member or leader of such a task force has been the most valuable management training opportunity I have encountered. In addition, it has also been among the most rewarding experiences.

You are called upon to bring your experience to bear upon a real and usually urgent problem. The task force recommendations benefit from the organization-specific experience of the members, i.e., they can be implemented. Task force membership is also a form of recognition which is difficult to match. (For a specific account of a task force experience, see Sect. 5.5.)

16.6 Summary

Among the short-term "General Management Training" programs available, we have identified the two ends of a spectrum:

1. The lecture format
2. The participative format

The basis for selection among a wide variety of courses is considered from the viewpoint of the Hi-Tech professional.

A particularly successful management development program is described. It lends credence to the potential of such an undertaking in a creative Hi-Tech environment.

A sample of a full-time academic program is presented.

The growing role and value of E-learning programs are noted.

In-house task force participation provides a most valuable learning experience for every level of management.

Recommended Further Reading

The following 14 recommendations are offered to the reader motivated to pursue further study. It is a very limited selection from a very extensive literature.

Not that other books lack relevant content regarding the topics covered. My selection criterion could best be expressed in terms of an "equivalent signal-to-noise ratio"—one that relates *relevant* content to the time invested in reading (relevant to Science and Hi-Tech).

Engines of Tomorrow: How the World's Best Companies Are Using Their Research Labs to Win the Future

This is a significant and comprehensive work that not only tracks the evolution of industrial research but details current practices at some of the world's best labs. I have not come across any other book about R&D on this scale—combining history, management issues and cutting-edge projects (*Robert Buderi, Simon and Schuster 2000, 600 pgs*).[1]

High Output Management

High on my list of relevant material is this most readable book by Andrew S. Grove, Former President of Intel First published by Random House in 1983, a second edition is available from Vintage Press, 1995, 230 pages (see Footnote 1).

---

[1] References to the "numbers of pages" are normalized to a standard page of 60 characters to the line and 30 lines per page.

R. Kay, *Managing Creativity in Science and Hi-Tech*,
DOI 10.1007/978-3-642-24635-7, © Springer-Verlag Berlin Heidelberg 2012

Advise to a Young Scientist

This small book by the late P.B. Medawar, 1960 British Nobel Laureate in Medicine and Physiology, is most rewarding for its style, humor and insight. He had the unique ability to offer advice without appearing to preach. Medawar projects the image of a great scientist with all the attributes of a well-balanced man of the world. A book I would recommend to any student entering graduate studies in science or engineering and to every practicing researcher every 5 years. Published by Basic Books, 1979, 106 pages.

Why Read Peter Drucker?

A thoughtful case is made for reading Peter Drucker, 60 years later, by A.M. Kantrow, HBR November 2009.

Alan M. Kantrow is a professor of management at the Moscow School of Management in Skolkovo.

Management

This classic by Peter Drucker is on the bookshelf of many practicing managers. It is more a reference book in which you read a chapter relevant to a particular issue than cover-to-cover reading matter. An excellent writer, Drucker has a lot to say and says it well.

First published in 1973, a revised edition is available from Harper Collins (2008), 715 pages (see Footnote 1).

Innovation and Entrepreneurship

One of Peter Drucker's later books; it has a chapter (75 pages) on the Practice of Entrepreneurship. It distills much of the, here relevant, experience of this most influential author, published by Elsevier 2007, 400 pages (see Footnote 1).

The Lean Startup

Based on the author's serial experience with software startups, this articulation of what it takes to manage a successful startup has found wide acceptance. One of the few authors in the field who sees value in sharing his failed approaches as well. Eric Ries (Fall 2011), Crown Business, Publishers.

Behavioral Science: Concepts and Management Applications

This is a condensed and at the same time very readable collection of the classic teachings of Behavioral Science applicable to management. It is based upon the writings of six of the most influential exponents (McGregor, Maslow, Herzberg, Likert, Argyris and Blake). It also includes some company experience which has been reported in the literature and which has influenced management practices the world over. The book also contains many references to the original material upon which the book is based.

The ideal place to start (RK), By Harold M.P. Rush, published by The Conference Board, 845 Third Avenue, New York, N.Y. 10022 in 1969 as Conference Board Report No. SPP 216. 435 pages (see Footnote 1). (Out of print, but well worth pursuing from out-of-print book vendors)

Work and the Nature of Man

Highly recommended to those who need to overcome whatever has deterred them from reading the work of a psychologist.

Eminently readable and stimulating as well as relevant to our topic. By Frederick Herzberg, published by the World Publishing Company, Cleveland, Ohio 1966, 200 pages.

Management of Organizational Behavior

Here the approach is that of the behavioral sciences. What makes this particular book commendable is that both authors have had relevant management experience at Bell Laboratories and in the Navy, respectively. The "Situational Leadership Theory," which has come out of the Center for Leadership Studies at Ohio University, becomes cumbersome in the attempt to make it serve as an umbrella for the entire subject. The book provides a good overview of the reported work relating to organizational behavior and leadership. By Paul Hersey and Kenneth H. Blanchard, published by Prentice-Hall, Inc., New Jersey 7632, 3rd Edition 1977, 450 pages (see Footnote 1).

Culture's Consequences: International Differences in Workrelated Values

Sociological study based on extensive data. Not specific to the Science and Hi-Tech environment, probably the most ambitious account of the international aspects of work-related values. This work has been updated and extended to many parts of the world. See http://www.geert-hofstede.com/geert_hofstede_resources.shtml by Geert Hofstede, Sage Publications, 1984, 480 pages (see Footnote 1).

Managing High Technology Companies

The emphasis here is on marketing, production, quality control, and finance. An experienced manager offers his views to mid-and top-level managers of Hi-Tech companies. The author runs the Executive Institute for the Management of High-Technology Companies sponsored by the American Electronics Association and Stanford University. By Henry E. Riggs, published by Van Nostrand Reinhold Company, N.Y, 1983, 400 pages (see Footnote 1).

In the Plex: How Google Thinks, Works and Shapes Our Lives

This is a comprehensive and very readable account of Google's history. Comprehensive insofar as it covers technical, business, legal and management issues which influenced Google's development. Readable in terms of succeeding to explain complex technical and financial matters as well as by providing some insight into the personalities involved. By Steven Levy (2011) Simon Schuster, 430 pages.

Who Says Elephants Can't Dance

A surprisingly readable account of the turn-around of IBM in the 1990s by the man who did it. Here you find an extreme example of what makes for success and failure in the world of Hi-Tech and develop a new appreciation for "strategic thinking." One of the great CEO's of our time. By Louis V. Gerstner, Jr., Harper Business, 2002.

Appendix

A.1 Management Initiative

The following relates to the discussion in Sect. 6.2.

Example of Five Levels of Management Initiative

Situation: Expectation of project review, date uncertain.

(The following are possible responses to the situation on the part of the project leader, corresponding to the five levels of initiative.)

Level 1: Project leader waits for her manager to announce date for review and what is wanted in the way of a presentation.

Level 2: Project leader asks manager for date and expectation.

> *Potential problem with levels 1 and 2: Project leader may get inadequate time to prepare and leaves the matter of content to her manager.*

Level 3: Project leader proposes content of presentation to her manager.

Level 4: Project leader consults with project members to plan presentation and informs her manager of what she expects to do.

Level 5: Project manager consults with project members, prepares presentation and sends copy to her manager.

> *This last approach gives the project leader maximum control of time for preparation, and content of presentation.*

A.2 Ranking Criteria for an R&D Function

(Relates to the discussion in Sect. 8.3.3.)

The difficulty of arriving at a satisfactory set of criteria for ranking creative people in a Hi-Tech environment was pointed out. The following represents such a set of criteria which has evolved over time.

R. Kay, *Managing Creativity in Science and Hi-Tech,*
DOI 10.1007/978-3-642-24635-7, © Springer-Verlag Berlin Heidelberg 2012

1. Technical Accomplishments—Quality and Quantity

   • Results—scientific or technical achievement
   • Creativity, novelty of ideas and new directions
   • Problem selection and approach
   • Initiative and resourcefulness
   • Level of independence
   • Technical breadth and depth

2. Internal Impact—Quality and Quantity

   • Effective contribution to group goals
   • Technology transfer
   • Patent activity
   • Reports and presentations
   • Strategic significance of work
   • Interaction with other parts of company
   • Consulting
   • Task force participation
   • Leadership and impact upon people
   • Recruiting
   • Awards

3. External Impact—Quality and Quantity

   • Honors and awards
   • Publications and presentations
   • Invited talks
   • University relations
   • Professional society activities
   • External funding

4. Administrative and People Management[1]

   • Impact of group
   • Recruiting
   • Setting direction and leadership
   • Resource allocation and collaboration
   • Budget control and funding
   • Communications
   • Performance planning, counseling, ranking
   • Equal opportunity
   • Safety
   • Security

---

[1]Primarily for managers.

A.3   New Venture Staffing Plan

Staffing Plan—Salary, Fringe and Burden (Year 1)

Function	Position	Salary Ann.	S, F & B Mos.	1	2	3	4	5	6	7	8	9	10	11	12	Total Year 1
G&A	CEO	120	26.00	26	26	26	26	26	26	26	26	26	26	26	26	312
	Admin/Off.	60	13.00	13	13	13	13	13	13	13	13	13	13	13	13	156
	Secty	40	8.67													0
Finance	VP (1/5)	40	8.67	9	9	9	9	9	9	9	9	9	9	9	9	104
	Acc'tng/Pur	50	10.83				11	11	11	11	11	11	11	11	11	97
	Total			48	48	48	59	59	59	59	59	59	59	59	59	670
Marketing	VP	120	26.00	26	26	26	26	26	26	26	26	26	26	26	26	312
	Sr. Mktng	120	26.00			26	26	26	26	26	26	26	26	26	26	260
	Sys. Eng	80	17.33											17	17	35
	Jr. Mktng	80	17.33													0
	Secty	36	7.80													0
	Total			26	26	52	52	52	52	52	52	52	52	69	69	607
Tech Ops	VP	120	26.00	26	26	26	26	26	26	26	26	26	26	26	26	312
	Serv. Mgr.	120	26.00	26	26	26	26	26	26	26	26	26	26	26	26	312
	EE/Progr.	100	21.67	22	22	22	22	22	22	22	22	22	22	22	22	260
	ME	100	21.67	22	22	22	22	22	22	22	22	22	22	22	22	260
	Tech	50	10.83			11	11	11	11	11	11	11	11	11	11	108
	Progr	60	13.00						13	13	13	13	13	13	13	91
	Secty	36	7.80						8	8	8	8	8	8	8	55
	Total			95	95	106	106	106	127	127	127	127	127	127	127	1398
Total				169	169	206	217	217	237	237	237	237	237	255	255	2674

G&A Gen. & Admin. Expense, *Fringe Benefits* 30% of Salary, *Burden Rate* 100% of Salary plus Fringe

References

Archibald, R.D. (2003): Management of High Technology. Programs and Projects 3rd Edition (JohnWiley & Sons, New York)

Ascheron, Claus and Kickuth, Angela (2004): Make Your Mark in Science: Creativity, Presenting, Publishing, and Patents, A Guide for Young Scientists, (JohnWiley & Sons, New York)

Backus, John: In "Research, Failure is the Partner of Success", Research Mgmt. Vol. 27, No.4, Jul-Aug 1984, pp. 26–29

Bardhan, A.D., Kroll C.A., (2006): Industry and Innovation, Vol. 13, No.1 69–95, March 2006. Competitiveness and an Emerging Sector: The Russian Software Industry and its Global Linkages. UCB Haas School of Business

Batelle report by M. Grueber, R & D Magazine, December 2010. http://www.battelle.org/aboutus/rd/2011.pdf

Baumol, W.J. (2008), Entrepreneurs, Inventors & the Growth of the Economy, Conference Board report #R-1441–09-RR

Brander, J. Du, Q. Hellmann, T. 2010 The Effects of Government-Sponsored Venture Capital: International Evidence. pg 27 http://sites.kauffman.org/efic/conference/BranderDuHellmannJune9th2010.pdf

Brandt, Steven C. (1986): Entrepreneuring in Established Companies (Dow Jones-Irwin, 1986 and Signet Books, 1987) p. 122

Brown, Deaver (1980): The Entrepreneur Guide (Ballantine, New York) p. ll

Buderi Robert (2000): Engines of Tomorrow, Simon & Schuster

David, Edward E. (1985): MIT Report, June 1985

Danielmeyer, H.G. (1997): European Review (1997), 5: 371–381; The development of the industrial society

Danielmeyer, H.G. & Takeda, Y. (1999): The Company of the Future. Springer Verlag

Deloitte and NVCA, 2010 Global Trends in Venture Capital: Outlook for the Future http://www.deloitte.com/assets/Dcom

Drucker, Peter (1985a): Management: Tasks, Responsibilities, Practices (Harper Colophon Books) pp. 649, 652

Drucker, Peter (1985b): Innovation and Entrepreneurship (Harper & Row, New York) p. 171

Drucker, Peter (1988): Leadership: More doing than dash'. Wall Street Jour, Jan,1988, p. 14

Drucker, Peter (1988): Best R&D is business-driven. Wall Street Jour. Feb. 10, 1988, p. 20

Drucker, Peter (2007): Innovation & Entrepreneurship, Elsevier

Feynman, Richard P. (1988): What Do You Care What Other People Think? (W.W. Norton)

Gerstner, Louis V. Jr., 2002: Who Says Elephants Can't Dance, Harper Business

Gomory, Ralph E. (1987): "Dominant science does not mean dominant products". Research & Development, Vol. 29, No. 11, pp. 72–74

R. Kay, *Managing Creativity in Science and Hi-Tech*,
DOI 10.1007/978-3-642-24635-7, © Springer-Verlag Berlin Heidelberg 2012

Google's Project Oxygen (2011), NY Times 3/13/2011, BU 1, 7

Grove, Andrew S. (1985a): High Output Management (Random House, New York) pp. 172–180

Harris, Gardiner, (2011): The People Skills Test, NY Times, 7/11/2011, p. 1

Hart, Gary L. (1977): "A workshop approach to improving material performance". Research Management, Vol. XX, No.5, Sept. 1977, p. 16

HBR (1986): Harvard Business Review, Ten-Year Index 1986

Herzberg, Frederick (1966): Work and the Nature of Man (World Publishing, Cleveland, Ohio) pp. 69–91

Hiltzig, Mathew (1999): Dealers of Lightning, History of Xerox PARC. Harper Collins

Hofstede, Geert (1980): Culture's Consequences: International Differences in Work related Values, Sage http://www.geert-hofstede.com/

Humboldt-Kosmos (2009): Cultures of Creativity: The Challenge of Scientific Innovation in Transnational Perspective Proceedings of the Third Forum on the Internationalization of Sciences and Humanities. November19–20, 2009 London

Kay, Ronald, (1991): Kreativitaet und Innovation, Beitrag zu Innovations - und Technologie - Management, p. 39, Verlag C.E. Poeschel, Stuttgart

Kerzner, H. (2009): Project Management -The System Approach to Planning, Scheduling and Control 10th Edition (John Wiley & Sons, New York)

Kets de Vries, M. (1985): "The dark side of entrepreneurship". Harvard Business Review, Nov.-Dec. 1985, p. 160

Kramer, A.E. (2010): Innovation, by Order of the Kremlin, New York Times, April 4, 2010

Lehrer, Mark (2000): [Has Germany Finally Fixed Its High-Tech Problem?: The Recent Boom in German Technology-Based Entrepreneurship Jul 01, 2000. Prod. #: CMR181-PDF-ENG

Leonard, Dorothy; Swap, Walter (1999): When Sparks Fly, Igniting Creativity in Groups, HBS Press 1999

Levinson, H. (1980): "Criteria for choosing chief executives". Harvard Business Review, July-August 1980, p. 113

Levinson Harry (2006): On The Psychology Of Leadership (Harvard Business Review Facebook)

Likert, Rensis (1967): The Human Organization: Its Management and Value (McGraw-Hill, New York)

Levy, Steven (2011): In The Plex of Google, Simon Schuster

Maslow, Abraham H. (1954): Motivation and Personality (Harper & Row, New York)

McGregor, Douglas (1960): The Human Side of Enterprise (McGraw:-Hill, New York)

Medawar, P.B. (1979): Advice to a Young Scientist (Harper & Row, New York)

MIT (1987): The MIT Management of Technology Program, The Alfred Sloan School of Management/The School of Engineering, MIT 1986–87

Moyers, Bill (1989): A World of Ideas (Doubleday, New York)

Oncken, William, Jr, Wass, Donald L. (1974): "Management time: Who's got the monkey?" Harvard Business Review, Nov.-Dec. 1974, p. 75

Peters, Thomas J., Waterman, Robert H., Jr. (1982): In Search of Excellence (Harper and Row, New York) p. 285

Porter, Michael & Kramer, Mark (1999): Philanthropy's New Agenda: Creating Value HBR Nov-Dec. 1999

J.B. Quinn, P. Anderson and S. Finkelstein (1996): Managing Professional Intellect: Making the Most of the Best Harvard Business Review, Mar-Apr 1996

Ries, Eric, (2011): The Lean Startup, Crown Business Publishers

Riggs, Henry E. (1983): Managing High-Technology Companies (Van Nostrand, New York)

Rush, Harold M.P. (1969a): Behavioral Science: Concepts and Management Application (The Conference Board) p. 19

Rush, Harold M.F. (1969b): Behavioral Science: Concepts and Management Application (The Conference Board) p. 21

Saxenian, A.L. (2002): Local and Global Networks of Immigrant Professionals in SiliconValley, Public Policy Institute of California

Scheer, A.W. (2010): Spiele der Manager, p. 13 (In German) IMC

Thomas, Uwe (2005): Stiefkind Wissenschaftsmanagement Friedrich Ebert Stiftung. (In German)

Uttal, Bro (1983): "The lab that ran away from Xerox". Fortune, Sept. 5, 1983 Wall Street Journal (1986): Feb. 28

Young, John (1983): In Center for Integrated Systems, ed. by F.H. Gardener, Hewlett-Packard Journal, Nov. 1983, p. 24

Wallace J. & Erickson J. (1992): Hard Drive, Bill Gates and the Making of Microsoft, John Wiley & Sons, Inc

Wang, Jennifer (2011): Entrepreneur Magazine - February 2011 Google's recruiter on how to build a superstar team

Wright, Gavin (2007): Historical Foundations of American Technology, Stanford University, Conference Board's Workshop *Perspectives on U.S. Innovation and Competitiveness*, Fig.5

Zaleznik, Abraham (1977): "Managers and leaders: Are they different?". Harvard Business Review, May-June 1977, p. 67

Zaytsev, Evgeny (2005): AmBAR. http://www.svod.org/files/resources/exec_summary_tutorial.pdf

Zaytsev, Evgeny (2011): Private Communication

Index